The Myth of Scientific Literacy

The Myth of Scientific Literacy

▲ MORRIS H. SHAMOS

Rutgers University Press
New Brunswick, New Jersey

Library of Congress Cataloging-in-Publication Data

Shamos, Morris H. (Morris Herbert), 1917–
    The myth of scientific literacy / Morris H. Shamos.
      p.   cm.
    Includes bibliographical references and index.
    ISBN 0-8135-2196-3
      1. Science—Study and teaching—United States.   2. Literacy—
United States. I. Title.
Q183.3.A1S45   1995
507.1′073—dc20   94-41057
                                             CIP

British Cataloging-in-Publication information available

To Marion—for her patience and support during the preparation of this book
and
To Joss and Alex—with the hope that during their lifetimes, the notion of scientific literacy may become more meaningful than it is today

▲ CONTENTS

FOREWORD   ix
by Mary Budd Rowe

PREFACE   xi

CHAPTER 1   1
A Crisis in Science Education?

CHAPTER 2   23
Public Understanding of Science

CHAPTER 3   45
The Nature of Science

CHAPTER 4   73
The Scientific Literacy Movement

CHAPTER 5   101
The "Two Cultures"—and a Third

CHAPTER 6   129
Recent Approaches to "Scientific Literacy"

CHAPTER 7   157
The Future of "Scientific Literacy"

CHAPTER 8 189
How Much Scientific Literacy Is Really Needed?

CHAPTER 9 215
*Conclusions—and the Road to a Possible
Solution*

EPILOGUE 229

NOTES 239

INDEX 253

▲ FOREWORD

▲ Morris Shamos is a scientist and educator to whom we should listen. He cares passionately about what happens to science education in the United States. *The Myth of Scientific Literacy* is not just the expression of an armchair critic. On the contrary, it reflects on decades of work and learning. In the early seventies, for example, when he was doing physics research and teaching at New York University, he started a science curriculum development project for elementary schools known as COPES. Among other things, Shamos tells us in this book what he learned from a decade or more of development and dissemination of that project. I took some part in the project as a developer and came to enjoy and respect his insistence on evaluating and learning from real data as a basis for making changes. In my experience, that was a rare trait among science curriculum developers.

For as long as I have known him, Morris Shamos has paid close attention to what was happening to education in the sciences at all academic levels—and he would get involved. He was one of a small group of scientists who took an active part in the National Science Teachers Association (NSTA), which at the time was an association primarily of secondary school science teachers. He believed NSTA could provide a forum where scientists and science teachers could work together fruitfully on behalf of science education. He served a term as president of that organization and has continued actively to support it and to challenge its pronouncements from time to time.

In this book Shamos argues compellingly against what popularizers of science (including public media specialists as well as science educators) are doing to public understanding of science under the guise of developing some ill-defined thing called scientific literacy. Because of his lifetime of experience as a scientist and teacher in academia, as chief scientific officer of a major international technological company, as a science curriculum developer and an activist on the national scene in science education, what he is saying deserves our attention. He distills all his experience as a scientist and educator into a brew that for many will taste bitter because it erodes away the underpinnings of the scientific literacy movement. Again, as so often in the past, he puts himself on the line by suggesting how we might start on a more productive agenda.

As always, I find his thinking stimulating, challenging, and productive. Let me warn you, however, Shamos is blunt. He did not set out to make us feel comfortable; quite the contrary. He did not set out to be popular. He cares too much about what happens to education in the sciences to let any of us settle for comfortable fads. Instead, he offers us a compelling invitation to inquiry. *The Myth of Scientific Literacy* has much to say in this era so dominated by pressures for national standards and accountability demands.

Mary Budd Rowe
Stanford University
Fall 1994

▲ PREFACE

▲   Science is usually portrayed to the public, along with mathematics, as being the most rational of all human enterprises. Yet the reasoned approach that supposedly characterizes the practice of science has not carried over very well into science education. There are a number of reasons for this: some are dictated by the customs of the science education community; others have to do with the role of science as seen by literate Americans; and still others are concerned with the nature of science itself. The problem is not new; since the early part of this century, spurred on largely by John Dewey's educational theories, the precollege science education movement in America gained momentum at an astonishing rate, matched only by the growth in the high school population itself. Next to the traditional 3R's, science soon became the most prominent subject area in the high school curriculum. Indeed, as measured by the time spent by secondary school teachers in the various subject areas, science presently ranks a close third, after English and mathematics. Thus, at least in terms of educational stature, science now appears to be on a par with reading, writing, and arithmetic.

Yet despite all the attention given to science education in this century, whether rationalized as a cultural or a practical imperative, beyond training a relatively small number of students for the science and engineering professions the science education movement has failed to penetrate the consciousness of the American public in a manner that even borders on what might be considered "scientific literacy." The fact is that regardless of all our efforts, by any reasonable measure we remain predominantly a nation of scientific illiterates. Moreover, the future prospects of achieving a common literacy in science seem equally dismal—so poor, in fact, that if assessed purely on a pragmatic cost/benefit basis the effort would surely be abandoned. But education cannot be measured this way, so that failure to achieve a given educational objective is generally countered with renewed effort, new ideas, and more money in the hope of evaluating and dealing with the causes of the failure; only in the extreme will the objective be thoroughly modified or perhaps even abandoned. In the case of scientific literacy, we shall see that the science education community, thoroughly imbued with what now appears to be no more than a utopian dream, has failed to assess this goal realistically and continues to brush aside suggestions that science education must seek a *radically* new approach if it wishes effectively to communicate science to the general public. And our politicians and the media have been equally indiscriminate in their advocacy of scientific literacy as something the schools could *easily* accomplish given the will (and pressure) to do so.

Public understanding of science, an idiom often used for scientific literacy, is perhaps more meaningful in this context because one can distinguish several "publics" as the target audiences for science understanding. First, there are the public interest groups dealing with science or technology-related issues; then come the legislators, whom the public interest groups seek to influence; and finally the largest segment, the general public, the informed and uninformed, interested and uninterested. Here, one might also add the science educators, an obviously highly motivated and concerned "public," and the scientific public—that is, those scientists whose understanding of science is limited to such very narrow regions of highly specialized disciplines that they often fail to view the enterprise in its overall social (or even scientific) perspective. Of all these groups, the lay public comprises the vast majority, some 90 to 95 percent of the total adult population. Science clearly has different meanings for each of these publics, as does the concept of scientific literacy, but such confusion is largely the making of the science and science education communities. We have simply failed to appreciate the different kinds of science understanding that are important to these publics.

World War II turned a spotlight on science, one that caught in its glare not only the scientific establishment, but by reflection the entire science education enterprise as well. The rate of growth of scientific knowledge since World War II has been staggering, so much so that the average individual's knowledge of science today is proportionately far less than at any time since science became part of the traditional school curriculum. Not only did the war's end usher in an era of "big science," but the postwar drive for reindustrialization forced the schools and colleges to gear up for increased numbers of science and engineering students. When little more than a decade later the nation was startled by the appearance of Sputnik and all that this small satellite implied, the science education community undertook a massive effort to restructure the science curriculum in the schools and to reformulate the purpose of such education. That such effort has failed to achieve its most prominent objective, *universal scientific literacy,* is the focal point of this book.

The book seeks to clarify the purpose of science education, to examine the history and meaning of scientific literacy, and to set out the reasons for our failure to achieve such literacy in the past—as well as the unlikelihood of our achieving it in the foreseeable future. It is not my intent to fault past efforts toward this goal, most of which were honestly and innovatively conceived, nor to engage in school bashing, as has become fashionable. I simply want to point out that having given the problem our best shot, so to speak, and failed, we must now seriously question the premise of general education in science, that is, of science education for the nonscience student. The book deals mainly with the central educational issues surrounding science that have faced our nation since the early part of the century, particularly since World War II, when the pace of activity in this area quickened noticeably. In it I will discuss such pervasive and fundamental issues as why we teach science, what science we should be teaching and to what students, and what some of the social and economic implications of science education are. And behind all these issues is the overriding question of scientific literacy for the masses—the elusive goal that has so often been used as the rationale both for compulsory science education in our schools and for curricular experimentation.

Preparation of this book was precipitated by the so-called crisis in science education that was declared early in the decade of the 1980s, when support for science education virtually ground to a halt under the Reagan administration. The details of this "crisis" will be discussed later, but it appeared to me at the time that the rule of reason, in which science normally places such great store, was being abandoned in a headlong but potentially futile quest for an ill-defined goal called "scientific literacy." With others, I became concerned that the science education community was unwittingly deluding both itself and the general public, and that unless the high-sounding platitudes were put aside and the purpose of science education was clearly spelled out in simple, practical terms, general education in science would lose much of its credibility. It was a remarkable period in the history of the science education movement, notable for demonstrating how effective an educational scare can be when coupled with national economic and social concerns.

The problem of general education in science reaches well beyond the current nationwide clamor for improving education generally. The failings of U.S. public education, as usually evidenced by declining test scores, and the alleged bankruptcy of cultural literacy in our students are larger problems that may encompass science education as well, but the overlap is only marginal. Science education, we shall see, has its own unique set of problems, its own failings, and its own conception of literacy. The promise of a *meaningful* public literacy in science is a myth. However good our intentions, we have tricked ourselves into believing that what is being done with science in our schools can lead to such literacy. The folly of this position is that not only do we lack agreement as to the meaning of scientific literacy, but more seriously, we also lack any *proven* means of achieving even the lowest level of science understanding in our educated adult population. Yet the posturing and the promises of things to come continue. At his famous education "summit" held in October 1989, President George Bush, who wished to become known as the "Education President," together with the state governors, proclaimed, as part of his "America 2000" plan for reversing the slide in U.S. education, the following:

- By the year 2000, U.S. students will be first in the world in science and mathematics achievement.
- By the year 2000, every adult American will be literate and will possess the knowledge and skills necessary to compete in a global economy and exercise the rights and responsibilities of citizenship.

Ambitious political prose, but was it really conceivable that what had never been achieved throughout the entire history of American science education—namely, adult literacy in science—could be accomplished in a single decade? Obviously not. Yet once the commitment was made at so high a level, it seemed likely that federal initiatives for science education in the decade of the 1990s would be driven by this unrealistic vision of a quick fix, and in all the excitement the opportunity for a deliberative review of the process might simply fade away. We shall see that this is pretty much what happened. While few educators really believe the projected timetable, curriculum reform is once again in full swing and the purse strings have been loosened for new experiments in science education.

The problem of scientific literacy cannot be condensed into a single issue—nor even to a few—but encompasses a broad range of questions, some seemingly unrelated but nevertheless bearing significantly on the central problem. It is not simply a question of whether such literacy would be desirable—of course it would. But we must ask (a) is it actually essential for the public good, (b) is it attainable with reasonable effort, and (c) what does it really mean to be literate in science? Is scientific literacy an educational problem, as most make it out to be; a social problem, which is more likely; or a combination of several discrete problems, which I believe is the case. To appreciate the problem fully, we should look not only at how and why *scientific literacy* became a late twentieth-century educational challenge, but we must also look at its beginnings. We will find that there is a curious parallel between some of the basic philosophical questions faced by science in its early days and those we face today in science education. These topics are covered in the first four chapters. Chapter 5 takes us back about forty years to the famed "two cultures" controversy and suggests that a third culture, a science counterculture, is emerging as still another roadblock to scientific literacy.

What is it about the nature of science that makes scientific literacy such an elusive goal? What can we learn from our past mistakes, both in science and science education? And what factors on the fringes of science—the countercultures, the pseudosciences, and outright flimflam—conspire to help defeat the goal? If the subject matter of the book at times seems to range far from the central issue, as in chapter 3, which deals with the nature of science itself, or the sections in chapter 5 that deal with postmodernist and counterculturist views of science, it is because I believe that these topics contribute in some important ways to an overall *sense* of the problem that confronts us. Some readers may find the material in chapter 3 a bit heavy, particularly where it borders on philosophy. Such matters, however, go to the core of understanding how and why the scientific enterprise works, why it differs so much from other disciplines, and why most students consider science a hard subject. Readers may skim over most of these topics without losing the major thrust of the chapter, but if they are curious about the underpinnings of science, they can find some of the answers here. In either event, it is important that the reader appreciate some of the more important factors that make scientific literacy so difficult to attain.

Chapter 6 reviews the major current curriculum efforts in science education, analyzing their probable impact on science education generally and their contribution to the goal of scientific literacy. Chapter 7 looks as far into the future as one dare project in today's highly volatile field of science education. Included here are such topics as how to measure scientific literacy, the drive for nationwide standards and assessments, informal education, and, most important of all, the key role of the teacher in this enterprise. Because total literacy in science is an impractical goal, chapter 8 steps back to examine the question of how much literacy is really needed; that is, what fraction of society, if somehow made literate in science, might substantially alter the way that society as a whole looks upon the scientific enterprise? Coupled with this is the fundamental question of the purpose of general education in science.

Everything is brought together in chapter 9, which takes off from the premise that scientific literacy, as normally perceived, is no more than wishful thinking, and proposes some alternative approaches. Chief among these is that as an educational goal "scientific literacy" should be replaced by "science awareness" or "science appreciation" and that the public *must* learn how to seek credible advice from experts, possibly through some form of "science court" concept, if it wishes to be brought into the science loop through the democratic process. Included is a proposed curriculum guide that, by leaning heavily on technology, is more responsive than our traditional science curricula to the perceived needs and interests of the general public. These proposals should help ease the dilemma, but they will not produce the utopia about which so many are still fantasizing—a civilization in which *all* educated men and women speak the language of science, as once prophesied by Jacob Bronowski. I believe that no reasonable amount of effort can achieve this end in the foreseeable future. Finally, the Epilogue offers a summary of where we are in the seemingly fruitless quest of scientific literacy, and it investigates the probable role of the federal government in science education. The latter promises to take on added importance in the next few years, as will the new trend toward privatization of our schools.

Throughout the book I have stressed *adult* literacy in science rather than student literacy alone. The reason is simple: if the purpose of general education in science is to produce a literate public that is somehow able to make use of such knowledge in a societal sense, it is mainly the adult segment of our population that is in a position to contribute, by virtue of position and standing, to the overall good of society. It is well known that most students who are not particularly interested in science, even those who nonetheless perform well in school science simply because they have learned how to be good students, forget the science they learn soon after graduation, if not immediately after completing their science courses. Hence one's good performance in school science is not a reliable indicator of subsequent literacy in science. What should be measured is the retention or reinforcement of such knowledge into adulthood, and here, we shall see, our track record is very poor indeed. It may be that there are no curriculum reforms, however promising they may seem at the time in terms of improved student performance, that can accomplish the objective of producing a literate adult population. If this is so, we clearly must reexamine the purpose of general education in science.

In a certain sense, writing this book has been emotionally trying for me, for its main thesis is not readily embraced by most science educators and contradicts a major part of my own life's work as a teacher. After spending many years trying to communicate the spirit of science to nonscience students, I do not find it easy to confess, especially to myself, that for the most part these efforts have been fruitless. Making it all the harder, playing an iconoclastic role does not win one many friends in a tradition-bound field. It is small consolation that on reflection many of my colleagues who labor in this vineyard—scientists and educators alike—find it equally barren. Fortunately, we all gain a measure of comfort from teaching the serious science-bound students, small though their numbers may be, who make the overall adventure worthwhile.

This book is intended primarily for the broad science education community—for science teachers and educators, school administrators and officials faced with the unenviable task of assigning priorities and funds to the different academic disciplines, and, finally, for community leaders, legislators, and their concerned constituents who must ultimately judge the effectiveness of U.S. education and provide for its support. My hope is that the book's conclusions will prove persuasive to the reader, or at least that the background data and supporting arguments may help focus future debate on the issues.

I have profited immensely from discussions with many colleagues in the science and science education communities. I hope that the book accurately reflects their many suggestions, including views that were often at variance with mine. A few of these helpful individuals deserve special mention. Early versions of the manuscript were read by J. Myron

Atkin and Mary Budd Rowe of the School of Education at Stanford University; both of them were very generous with their time and helpful comments. I am particularly grateful to Professor Atkin for reinforcing, through his writings, my position on the proper roles of science and technology in a modern science curriculum, and for directing me toward the interesting fragment of educational history relating to Edward Teller (chapter 8). Professor Rowe, who also reviewed the final version of the manuscript, brought a different perspective to the work, namely, her considerable experience with the problems of preparing teachers of young children, the one area where the United States is particularly weak in science education. Beyond this, her broad knowledge of the overall spectrum of science education, as reflected in her comments and suggestions, proved invaluable in refining parts of the manuscript. I am also indebted to Clifford Swartz and Lester Paldy, of SUNY, Stony Brook, for their helpful comments on the manuscript, as seen from a combined science/science journalist viewpoint, each being both physicist and journal editor. Finally, my deep appreciation goes to Arthur Kantrowitz of Dartmouth College, who shared with me his current thoughts on the "Science Court" concept he proposed a quarter century ago, one that deserves a revival in the current pressing need for expert advice in science.

During the past few years I had the opportunity of participating regularly in the Scientific Literacy Seminar organized by Samuel Devons at Columbia University. Belying their title, these sessions simply confirmed my conviction that scientific literacy is indeed a myth. I am greatly obliged to Professor Devons and other seminar participants for many spirited and enlightening discussions on the subject.

The Myth of Scientific Literacy

CHAPTER 1

A Crisis in Science Education?

Good sense, which only is the gift of Heaven,
And though no science, fairly worth the seven.
—Alexander Pope (*Moral Essays*, 1731)

▲ The twentieth century has seen three major curriculum reform movements in U.S. precollege science education. The first, identified mainly with John Dewey's attempt to develop in students what he termed "scientific habits of the mind," began about 1910 and lasted until the post–World War II period. As we shall see, Dewey and his numerous disciples never succeeded in developing such critical habits of the mind, but their efforts did result in science becoming a part of the general education of most high school students. The second, beginning in the late 1940s and extending to about 1980, was triggered by the postwar drive for reindustrialization and subsequently spurred on by the embarrassment of Sputnik. Its major thrust was that U.S. science education was sorely in need of modernization and that science curricula and textbooks should better reflect the discipline as seen through the eyes of professional (academic) scientists. This was a truly massive curriculum reform effort, financed largely with government funds totaling more than two billion dollars. But it seemed to run out of steam in the late 1970s with the realization that while the reform was intellectually satisfying both to the scientific community and to science-bound students, and to a lesser degree to the science education community, its effect on the general student was at best marginal. The final and current movement began in the 1980s, and if the past is any guide, presumably it will continue for a school generation or two before its effect can be fully assessed. This movement has come to be known as the era of "scientific literacy" or public understanding of science, since its professed objective was to achieve this mental state in the population as a whole. In retrospect, the three movements may be seen as having been initiated by different constituencies: the first was directed largely by science educators; the second by academic scientists; and the third, taking place now, appears to provide for a much greater involvement of classroom science teachers than either of the other two, although science educators and, in a minor way, academic scientists are also involved. It almost seems as though the different groups entered upon the scene in the belief that each might do better than its predecessor.

▲   A SEASON OF SELF-APPRAISAL

The decade of the 1980s is perhaps best characterized as one of trial and turmoil in the education arena generally, particularly in science education. The heady days following World War II, when science education was in its golden age of government support and seemingly endless curriculum reform, turned to pessimism in the late '70s, when questions were raised regarding the effectiveness of the new programs. Support for science education came to an abrupt halt in the early '80s. By the end of the decade it was once again on the rise, presumably reflecting the hope of science educators (and the Congress) that enough could be learned from past failures to avoid making the same mistakes in the future—an expectation apparently based more on optimism than on sound premises. The '80s marked a turning point in the affairs and fortunes of science education. The movement that began with a bang at midcentury had pretty much fizzled out by the beginning of the decade for lack of direction and continuing support, and the science education community floundered in a sea of uncertainty. The most notable event of the decade was probably the discovery that to focus attention on a problem, real or perceived, it was best to label it a "crisis." Such was the case with science education.

The "crisis" was pronounced in October 1982, when a select Committee of the National Science Board issued its initial report assessing the state of precollege education in mathematics, science, and technology in the United States. Dramatically titled *Today's Problems, Tomorrow's Crises*,[1] the report followed a convocation held earlier that year on the same subject under the joint sponsorship of the National Academies of Science and Engineering, at which Margaret Heckler, then a congressional representative from Massachusetts, spoke of an "education and manpower training crisis" facing the nation. As might have been expected, her characterization was contagious. The conferees were quick to agree that a "crisis" existed, the commission adopted the description for its initial report, the news media pounced on it, and a massive ripple effect began spreading through the nation.

The reasons given for the "crisis"? A dearth of qualified science teachers, a lack of interest among students in careers in science, and declining test scores—all of which threatened to cause shortages of professional scientists and engineers. Worse yet was the alarming conclusion that unless all citizens became scientifically literate they would be unable to function properly in the workplace, or even as intelligent voters and consumers. Before long, it was feared by some, we could become a second-rate nation.

The report became a rallying cry for the science education community, which proceeded to vent all kinds of grievances under the "crisis" banner. Dozens of reports soon followed from state and local education groups, as well as from many foundations and scientific organizations, all supporting the earlier report and all carrying the same note of urgency. So numerous were the reports that Fred Hechinger, then education editor of the *New York Times*, was prompted to observe in one of his columns: "Conservatives say that you cannot solve the schools' problems by throwing money at them. But can you solve them by throwing reports at them?"[2] This was only a year after the commission's initial report was issued and just a few weeks after its final report. At that time there was only a handful of reports on the subject, including another widely read study, *A Nation at Risk*,[3] released a few months earlier; but five years later there were more than three hundred, and reports deploring the state of science education continue to this day.

The so-called crisis quickly took on a life of its own as numerous nonscientific groups and legislators at all levels of government added their voices to what appeared to be a real concern. But it was not all lip service. A number of states and local communities, fearing the worst, took direct action by mandating more science and mathematics in their schools. Most increased their salaries and incentives for teachers, a measure that must be counted as the single most productive response to the cry of "crisis." Science-oriented programs and activities outside the classroom were initiated in many communities and school systems. The news media began showing renewed interest in science and education. Several new science magazines aimed at the lay reader were launched (most of which have since failed). There seemed to be no end to how far the movement would spread.

Such was the genesis of the "crisis." Suddenly, a few years later, the crisis was over. It simply ran out of steam. "What was a clamor in 1982 has diminished to a murmur in 1986," reported Ward Worthy of the *Chemical and Engineering News*, who, in the July 19, 1982, issue of that journal, described what he called "convincing evidence of crisis status," yet reported in the March 10, 1986, issue that "the crisis in precollege science and mathematics—so much in the news just four years ago—is over." Worthy, an experienced observer of the educational scene, must have been more surprised than most by this turn of events, for he went on to say, "No, the problems that provoked the crisis haven't been solved. . . . Nevertheless, the crisis, as measured by the benchmarks of political and media attention, has vanished." So was the "crisis" all smoke or was there some fire as well?

▲ THE REAL CRISIS

Actually, there *was* a crisis of sorts, but not the one depicted by those raising the hue and cry. The real crisis was only indirectly related to the state of science education, teacher shortages, or scientific illiteracy, or to the many other faults perceived by the science education community. Yet in the midst of all the finger-pointing, rhetoric, and general confusion, it somehow escaped much notice.

The real crisis had to do with the climate of concern for science education brought on by the actions of the new Reagan administration soon after it took office in 1981. It announced that it intended to dismantle both the National Science Foundation's (NSF) Science Education Directorate, which was the major source of support for experimental programs in science education, and the newly established Department of Education. These were serious threats to the science education community, which for many years had looked to both agencies, particularly to the NSF, for support of its projects, and science educators fought back with the one weapon they could muster at the time: a cry of "crisis." In the end, both these case histories of the educational ineptitude of the Reagan administration were defeated by intense lobbying efforts, and the "crisis" quietly slipped away, but only after the Science Education Directorate had been put out of business for two years. Despite the demise of the directorate, the "crisis" prompted renewed interest in the perceived problems of science education, including the perennial issue of scientific literacy for all Americans.

It was no coincidence that a crisis was declared at about the same time as federal support for science education dried up, and that it disappeared once funding was restored and the NSF became active again in precollege science education. *This* was the only crisis; the clamor over other issues, while passionately believed by some science educators, was mostly rhetoric put forth by special-interest groups, some beneficially, others hoping to benefit from the expected wave of renewed federal funding once Congress moved to restore the Reagan administration's budget cuts. Worthy was right; the problems that were cited as precipitating the "crisis" had not been solved. Perhaps more to the point, not only are these problems much the same today as they were at the onset of the "crisis," but they are also the same ones that existed a quarter century earlier when Sputnik was launched. We don't hear as much about them now only because the science education community, perhaps reassured by all the clamor and the restoration of funds for support of science education that someone in Washington cares about them, seems to have adopted a wait-and-see attitude.

Soon after the "crisis" ended, Audrey Champagne, then a senior program director of the American Association for the Advancement of Science (AAAS), commented on its political nature:

> The perception of crisis is partially a product of hyperbole employed in the many national education reports. In part, however, the crisis is the product of the Reagan administration's need to demonstrate its interest in education. The administration early on ignored education. . . . Thus, when the national reports began to emerge, the administration seized the opportunity afforded by the reports and loudly proclaimed the need for improved science and mathematics education. The developing crisis was nurtured by special interest groups that potentially benefited from federal recognition of a crisis. The promise of federal dollars motivated associations of teachers, educational researchers, and social and natural scientists to contribute rhetoric consistent with the crisis mentality, thus further nurturing it. Is there really a crisis in science and math education? No, not really. The rhetoric notwithstanding, there is no reason to believe that the national security, economy, democratic way of life and science prominence are threatened by the low level of scientific and mathematical literacy in the general population.[4]

The full story has been told elsewhere,[5] but of all the rhetoric that characterized the "crisis" probably the most misleading was that relating to the purpose of universal (compulsory) science education, that is, to the *need* for widespread scientific literacy in our society and the prospects of achieving it through normal schooling.

Here we must distinguish between desirability and necessity. One can hardly fault the science education community for seeking to promote its discipline, for asserting that all educated individuals should know something about science, at least as an educational imperative if not for pragmatic reasons; and it had every right to deplore the cavalier attitude of the Reagan administration toward education generally and science education in particular. The same should be true of all academic disciplines, many of which voiced their concerns at the time over the misguided educational philosophy of the executive branch. But this was different. Here the science education sector was asserting that unless massive support (both moral and financial) were provided at once for science education, dire practical consequences could result for science and for the nation. While such warnings of "the sky is falling" managed to reverse the fortunes of science education, mainly by way of forcing reactivation of NSF's Science Education Directorate and producing substantial financial support from Congress for its programs, the arguments presented rested on shaky evidence then and have not become any stronger with time.

▲ THE "MANPOWER" NUMBERS GAME

The grounds for demanding greater support for science education, as pointed out above, were threefold: (1) a shortage of qualified science teachers; (2) decreased interest among high school students in careers in science, with the clear implication of a future shortage of scientists and engineers; and (3) declining test scores, both nationally and in international comparisons. As for the alleged teacher shortage, it turns out that this was based largely upon isolated anecdotal evidence, for on a national level it had no basis in fact. In a November 1983 survey of teacher shortages (published in 1985) it was found that nationwide only 2.5 percent of those teaching science in our public high schools were not fully certified by their states for the specific fields in which they were teaching (ranging from 1.6 percent for chemistry to 2.8 percent for physics).[6] Similarly, some 2.4 percent were teaching mathematics and 3.9 percent were teaching computer science with only temporary or emergency certifications. To put the matter into proper perspective, these numbers might be compared with the 2 percent in foreign languages, 3.5 percent in vocational education, and 12 percent in bilingual education who were teaching out of license.

Even this fails to tell the full story. Perhaps more to the point are the actual candidate shortages in these fields for that year. Of about 112,000 full-time teaching positions in science, only 180 went unfilled, and of 152,000 positions in mathematics, including computer science, 270 were unfilled. Any large organization with this number of employees, whether in government or private industry, would consider itself very fortunate if at any given time only one or two positions were vacant for every one thousand employees. Calling this a *critical* shortage of science teachers, when viewed in the overall picture, was at best unwarranted. Bear in mind that after teachers of English, who comprise fully one-quarter of our high school teaching staffs, come teachers of mathematics with 16 percent and of science with 12 percent. Following these are social studies with 11 percent, and then all other disciplines with much smaller faculties. Surveys of high school principals invariably show that they find it more difficult to fill vacancies in science with fully qualified teachers than in other disciplines, with physics vacancies being the most difficult to fill and biology the least. But somehow they manage to fill them—sometimes by extending their recruiting efforts beyond their immediate geographical areas; sometimes through job mobility, that is, by reassigning teachers who may be certified in more than one discipline; sometimes by pay differentials; and sometimes by recruiting active or retired scientists from industry.

One could question, of course, whether "certified" means qualified in all cases, and of course it does not. As in all fields, some science teachers are clearly better than others, either by virtue of training or experience, or because of their dedication to teaching. Some states are less demanding than others in their certification requirements, but barring a national system of certification—which many hope will one day come about—individual state certifications are the only guide available, and we can do no more than accept these as the guideline when assessing teacher shortages objectively. We have addressed only the problem of high school science teachers, who must be certified not simply in science but in one or more of the separate subject areas in science. The elementary schools, where teachers are generally far less qualified to teach science than any of the many other subjects they must cover, represent a vast wasteland (and a great opportunity), we shall see, in which to sow the early seeds of science.

▲ PRECOLLEGE SCIENCE ACHIEVEMENT

The issue of declining test scores is a vexing problem. Here one must distinguish between national assessments of science achievement over time, which may provide useful information on trends in science education and show up differences among school districts, and the international comparisons that draw so much public attention. The international comparisons, we shall see, can be very misleading as indicators of a nation's success in science and engineering, yet because of our consuming interest in international competitiveness on the economic and political levels, such comparisons seem to capture most of the attention of the American public. Shortly before the National Science Board issued its initial "crisis" report in 1982, the U.S. Department of Education published one of its periodic "science report cards," showing a small relative decline in student performance in science since its previous study five years earlier. But more to the point, in absolute terms, while student knowledge of science facts was high, its understanding of science was distressingly low.[7] This was its fourth such assessment. Interestingly, the fifth report card, on its 1986 study (issued in 1988), showed a small improvement in most areas of science tested, but overall student performance remained very poor, as it did on the sixth report card in 1990. Since the first such report, in 1969, the scores of seventeen-year-olds declined steadily until 1982, when they began to rise slowly but have yet to reach the 1969 level.

Average Scholastic Aptitude Test (SAT) scores of college-bound high school seniors had also been dropping steadily in the two decades preceding the "crisis" report, more so in the verbal portion than in mathematics. The latter was taken as further evidence that student interest in science had declined since the immediate post-Sputnik era, when the science education community was in its most active phase of curriculum reform. The conclusion was undoubtedly correct; interest in science did decline from its high point reached at midcentury. But this result had more to do with factors outside the educational system —that is, with social and economic forces that propelled more students into other disciplines and alienated some from science —than with poor curricula or teaching. It may be noted that in the same period, the American College Testing Program (ACT) showed a decline in mean scores in mathematics but a small rise in the natural sciences. The obvious interpretation of this is that the ACT program targets more specifically those students with a declared interest in science than does the math portion of the SAT, which is designed to test aptitude rather than achievement.

International comparisons of science achievement at the 5th, 9th, and 12th grade levels, in which the United States generally does poorly compared to most other nations, are also taken by some as an indicator of the declining interest and ability in science of U.S. students.[8] This conclusion, while possibly correct, is by no means obvious, for it fails to take into account several social and educational factors in those countries that may easily account for such a difference.

First, in all other industrialized nations, the competition among students for admission to a college or university is much keener than in the United States because fewer places are available. We are the only nation in the world that through its state university systems enjoys what might be termed "universal higher education." Any student who graduates from a high school is virtually assured of admission to some college or university, although not necessarily the student's first preference—provided the student can find the necessary financial support. Roughly 60 percent of our high school graduates (or about 50 percent of those entering high school) go on to college, with about half going to two-year colleges. Contrast this with the 30 percent going to college in Japan and the still smaller percentages in such industrialized countries as France, Germany, the former Soviet Union (FSU), and the United Kingdom. An important difference, of course, is that a much higher proportion of Japanese college students elect science and engineering as their major fields than do their U.S. counterparts. The net result is that Japan presently produces greater numbers of scientists and engineers than the United States, a fact that does not go unnoticed by those concerned about our international competitiveness. Another important difference is that the dropout rate of high school students is much lower in Japan than in the United States, leaving a larger pool (in percentage of school population) competing for the available college openings. It is noteworthy, incidentally, that with the recent collapse of its political and economic systems, the FSU's science education has fallen upon very hard times. Some 35 percent of all its high school teachers have left the education sector for other pursuits, leaving a huge gap and decreasing the number of gifted students studying science at the university level.

Thus, the incentive to perform in high school science and mathematics, once considered important benchmarks for college admission in the United States, appears to have diminished in the postwar period. The fact that poor achievement in science is no barrier to college admission or to earning a college degree hardly serves as an inducement to regard seriously this segment of the school or college curriculum. On the other hand, in countries where students must compete strenuously for admission to college, demonstrated proficiency in science is taken for granted as a prerequisite and high school students prepare themselves accordingly. But this alone should not account for the relatively weak performance of students in advanced placement courses, who are presumably more committed to science than the general student. Here, if the comparisons are to be meaningful, one must take into account not only the amount of science to which the cohorts in the various countries have been exposed by the time of testing, but also the kind of science that was emphasized—for example, facts, theory, or drill in problem solving—and the learning environments in school and at home. Objective testing must be done, of course; there is no easy alternative. But so many factors can influence competitive testing that careful selection of the test sample is vital if meaningful decisions are to be based upon such analyses. The least doubtful conclusions clearly derive from testing a single class with a single teacher; the next from interclass and interschool comparisons; and the most questionable from the cross-cultural test samples that are used in international studies.

Let us assume for the moment that the declining interest in science over the past two decades was not simply a correction to the rapid postwar growth in U.S. science and engineering, about which more will be said later, but a result of genuine disenchantment among high school students with possible careers in science. The question is whether this justifies a "crisis" atmosphere, and the obvious answer is that it is a crisis only if such decreased interest can be translated into *actual* shortages of scientists and engineers that will have a negative impact on the nation's educational, industrial, and military enterprises. Here is where the evidence is weakest and the greatest uncertainty persists. Many groups, governmental and private, have over the years routinely studied the nation's employment and demographic statistics and provided estimates of future scientific labor force needs, an exercise that has turned out to be anything but an accurate science. Chief among the federal agencies performing such studies are the National Science Foundation, noted particularly for its Science and Engineering Pipeline reports, the Census Bureau, and the Bureau of Labor Statistics; in the private sector, the Commission on Professionals in Science and Technology. All predictions of future shortages of scientists and engineers rest upon forecasts of supply and demand. The supply side is easier to estimate, being based on well-known school population statistics and past experience on how many beginning high school students eventually end up with baccalaureates in science or engineering—typically about 5 percent. But this is really a *gross* number. It fails to account for those who do not enter the professional work force after their science or engineering training, or those who enter a science or engineering field other than that for which they had been trained. Such job mobility—for example, physicists taking jobs as engineers or vice versa—while rare in universities, is well known in industry, particularly to fill spot shortages that sometimes occur in a given field within a company. All these studies pointed to a critical shortage of scientists and engineers by the turn of the century, variously estimated from a low of 100,000 to as many as 700,000. Such forecasts take into account the expected decline in high school graduates in the first half of the 1990s (by as much as 12 percent), which would show up as a decrease in college graduates by the end of the decade, all other factors remaining constant. However, the number of high school graduates then rises again and will even exceed current levels by the end of the decade.

The huge variance alone was very suspicious and suggested a basic problem in the design of such studies. The only sensible conclusion one could reach after looking at the postwar history of our supply of scientists and engineers was that the econometric model used by the NSF was flawed, more likely on the demand side, which is very difficult to forecast, than on the supply. As any labor economist knows, a model that ignores the effect of demand on the supply of labor forces must be regarded with suspicion. In July 1990 the National Research Council, having commissioned a comprehensive study that was published a year earlier,[9] confirmed that there were fundamental defects in the manner of arriving at the data needed to make these projections. This threw into doubt *all* the estimates of scientific labor shortages reached by the NSF. The report showed major discrepancies between the workforce estimates used by the NSF and those cited by the Bureau of Labor Statistics, and identified the sources of these variances. The predicted shortage was further disputed in a study presented by Robert White, president of the National Academy of Engineering,[10] which showed an actual surplus of scientists, and another by Alan Fechter,[11] also disputing the alleged shortage of scientists. And many others, including S. Allen Heininger, then president of the American Chemical Society and formerly vice president for research of Monsanto, failed to see a critical shortage of scientists and engineers on the horizon, certainly not as far as the needs of industry were concerned. Yet, despite mounting evidence that the forecasts were questionable, some high-level policy leaders continued to ring the alarm, including Richard Atkinson, chancellor of the University of California, San Diego, in his 1990 AAAS presidential address.[12]

The question of the size of the scientific work force turns out to revolve partly on definitions and partly on who is looking at the nation's needs. First, what constitutes a practicing scientist or engineer—having a degree in the field or working in the field, or both? And in the effort to track those high school and college students who, after expressing a genuine interest in science, turn away from it, one must deal with the definition of a science-bound student. Should this be based solely on the student's declaration (which is usually the case) or upon more objective criteria? The latter is particularly important because of the large attrition claimed by the NSF in the "science-bound" student population between the sophomore year of high school (10th grade) and college graduation (from the 20 percent of sophomores claiming interest in science to the 5 percent of that population actually earning college degrees in the field). How much of this apparent "loss" can be attributed to a casual rather than a serious "interest" in science remains an open question.

The academic sector sees the problem quite differently than industry. Colleges and universities need to attract sufficient numbers of the best-qualified Ph.D.-level scientists and engineers to fill the faculty positions that will become available over the next few years. This is further discussed below, but clearly any serious shortage of Ph.D.-level scientists and engineers is likely to be felt more keenly by universities than by industry, particularly in terms of the salaries required to attract such individuals to academic careers. The issue is not entirely economic, however, for there is a class of individuals that prefers the relative freedom of academic life, and given the opportunity would prefer an academic position to one in industry. In any event, the debate on future scientific work-force shortages does seem to be drawn somewhat along "party lines." We shall see that while the education sector (secondary school and college levels), on the strength of forecasts coming from so prestigious an agency as the NSF, kept insisting that a shortage was just over the horizon, the major consumer of scientists and engineers, namely industry, did not draw a like interpretation from the same work-force data. But students are not easily misled by such reports. There is an effective underground feedback system in our high schools and colleges that seems to inform students when certain fields are falling short of scientists and engineers, and also when there is an impending surplus. And students respond accordingly, so that with a time lag of several years most such peaks and valleys tend to even out. Moreover, one cannot ignore the effect of peer pressure on students' career choices, or even on their course selections.

▲ THE ROLE OF INDUSTRY

Most likely, after this fiasco, the methods of analysis will be refined, and more reliable labor statistics will be available in the future. But labor forecasting alone does not tell the full story; what actually happens is more meaningful. While the science education community, on the basis of the NSF forecasts, was warning of shortages, industry, which, after all, would be hardest hit by such shortages, offered little active support to the issue. There were some isolated expressions of concern. A few corporate executives urged that the work force generally ought to be better educated, and some companies even formed alliances with local school districts to provide funds and technical staff to assist in their science teaching programs. Yet there was no great clamor for a national effort to prevent the predicted shortage. Why? Presumably because industry does not react aggressively until it actually feels some real pressure, and there were no obvious signs that the threatened shortage was fast approaching, or even, as many executives believed, that it was approaching at all. Industry has seen spot shortages of scientists and engineers in the past, and will continue to see them in the future, simply because the supply of highly trained scientists and engineers cannot be turned on and off quickly to keep pace with a fluctuating demand. But the same is true of surpluses in these fields: there have been times in the postwar period when engineers in some industries were reduced to pumping gas or driving cabs. The economic recession of the early 1990s, which put many scientists and engineers out of work, is the most recent example of how risky long-range manpower projections can be.

American industry knows from experience that somehow it generally manages to fill gaps in its science and engineering work-force pools. This is done either through the sort of job mobility referred to earlier, including the use of what might normally be considered less qualified personnel—for example, having technicians do some of the tasks traditionally performed by scientists or engineers—or by importing foreign scientists. It also knows that as a last resort it can always attract the technical personnel it needs by making the demand side more attractive to them. In short, industry generally takes the view that should a shortage occur, market forces will attract more students into science to fill the gap. This position is amply supported by the supply and demand history of the technical employment market since World War II; somehow, perceived shortages were generally overcome, either through job mobility in the short term or in the long run by attracting more students through credible assurances of real opportunities in the field. The only time that the effect of market forces is not evident is when the market is not *seriously* exercised over the problem. Should a shortage ever become truly critical, industry can be expected to respond by providing the necessary incentives (financial and otherwise) to resolve the problem. Evidently some labor forecasters believe that market forces alone will not suffice, but both history and current experience do not support this view.

Initiative on the part of prospective employers of scientists and engineers appears to be a part of the equation that has not received the attention it deserves. The main problem with relying too heavily on demographics to forecast shortages in science and engineering is that it invites the simplistic solution of increasing the number of high school students entering the pipeline to ensure a greater output of science-related professionals at the end. Most "solutions" to a supposed shortage revolve about increasing student interest in science at the high school level and/or seeking ways to maintain that interest through college. Considering that if the NSF statistics were correct (which, as we have seen, is not the case for their long-range projections of shortages but should be true with respect to student demographics) fully three-quarters of the students who profess an interest in science as high school sophomores fall out by the time they complete college. Hence, clearly the "shotgun" approach of pushing more students into the pipeline makes little sense. Even if only 10 percent of students have a *serious* interest in science at some point in their high school careers, which I believe is more nearly correct than the 20 percent figure offered by the NSF, it should be obvious that increasing the retention rate of serious students ought to be far more effective than simply increasing the initial supply. But even increasing the retention rate of students already in the pipeline is not the most effective means. More effective still would be a retention program aimed at those leaving the pipeline, namely, those graduating with degrees in the natural sciences or engineering. According to the NSF's "Pipeline" report, one in five of these graduates enters a field other than those in which the degree was earned.[13] If correct, this means that there exists a *trained* pool of individuals equal to 25 percent of those actually entering employment in these fields, who for one reason or another have become disenchanted with science or engineering, or have found some other careers more attractive. Surely it makes more sense to work on retaining a portion of this group than to look for other solutions to a projected shortage of scientists and engineers, especially as this could be done in the short term since the group is already trained. This would be the obvious solution used by a resourceful industry once it believes that a critical shortage is truly on the horizon. Simply make the career opportunities in science and engineering more attractive, both financially and professionally, and industry would have little difficulty persuading many of those who would otherwise leave to continue on their original career paths. If only half of this group could be induced to remain in science, it would mean an increase of about 12.5 percent annually (i.e., about 20,000) in the pool of individuals ready to enter the scientific work force. Whether employment would be readily available for such an increased number each year is another matter that I will touch on later. The mere fact that this is not happening now should be evidence enough of the play of the marketplace, for it means that industry and government, the major employers of science and engineering baccalaureates, do not presently feel the need to provide such inducements; or put another way, these employers find the current supply more than adequate, even if short in some areas. That they *will* do so (particularly industry), if faced with a pressing labor shortage, should be taken as a foregone conclusion in a free economy.

As Robert Reich, of Harvard's John F. Kennedy School of Government, pointed out before becoming secretary of labor in the Clinton cabinet, when corporate America is unable to find enough skilled workers locally, it finds them elsewhere, either by sending the work offshore or by importing foreigners to fill the gap.[14] If need be, industry could extend this practice to fill most shortages that might develop in its science and engineering work force, as it already does to some extent— probably at lower cost than underwriting the science and technology education of native Americans as a recruitment tool. Reich adds that the highly publicized corporate support of education goes mostly for "in-house" training of selected employees, with very little going into public education at the primary and secondary school levels. For example, of the $2.3 billion that U.S. corporations contributed to education in 1989, only $400 million was earmarked for public schools— about 0.2 percent of total public school expenditures at the precollege level. Moreover, the often-heard argument that industry contributes much more to education through local taxes is somewhat deceptive, according to Reich, since most industrial plants benefit from local tax incentives as a means of attracting and keeping them in given communities. In many cases, in fact, it appears that industry benefits more from tax abatements it receives from the community than it returns in the way of support for education.

There is an ethical issue here as well. Should we encourage many more students to go into these fields? The question is especially important for minority students and women, who are presently underrepresented in science and engineering and therefore present both an untapped human resource and an opportunity to redress the equality issue that is often invoked. The projected (though faulty) shortage of scientists and engineers appeared to provide, among other things, a real opportunity to increase the roles of women and minorities in science and engineering. While no longer valid, this rationale is invoked even now by many advocates of such increased representation. The objective is laudable, but unfortunately the premise is defective. Indeed, should we seek to increase the work-force pool at all in these fields? Do we have unambiguous evidence of an impending shortage ten, fifteen, or twenty years from now, or should we rely upon the optimistic belief that career opportunities will be available for all —that a surplus is better for society than a shortage? Some believe that an oversupply might stimulate an increased demand, that high-technology industry might expand purely on the strength of the increased availability of technically skilled labor. But except in those nations undergoing rapid industrial growth, this is not the way a free economy generally works, especially in the U.S., where surplus professional employees who are otherwise competent are more easily dismissed than in most other free world industrialized nations. This means that if the U.S. were to encourage more students to aim for careers in science and engineering, employers would simply pick the best among them, leaving the others to find employment outside their fields. A large surplus of scientists and engineers may benefit the employer, but not those who will have spent years in training for what may then turn out to be low-salaried or highly competitive positions.

Career decisions usually last a lifetime and should not be dealt with lightly, nor be guided solely by apparent short-term needs of the marketplace. And science is a demanding discipline, requiring a dedication that transcends employment opportunities. Some contend that training in science or engineering prepares the individual to enter many other fields as well; hence, creating a manpower surplus would not necessarily prove a disadvantage to those who cannot find employment in their specific fields of training. One would like to think this true, but like most educational lore there is no reliable evidence for it and one could hardly advise students to go into science with this as a fallback position.

Few industries and fewer nations can afford the luxury of "stockpiling" scientists and engineers. It would therefore be irresponsible for the educational community to train many more scientists and engineers than the economy can reasonably absorb, just as it was reckless to put forth as a reason for the "crisis" the need to increase our scientific work force without accurately quantifying the problem. Unfortunately, the NSF's and similar forecasts of labor shortages added high-grade fuel to the flame, prompting many educators and politicians to climb on the "crisis" bandwagon. In the end, this served only to harm the *legitimate* claims of the science education community. Human resources in science and engineering are not education-limited, but demand-limited; unfortunately, the supply cannot be turned on and off as quickly as the demand may sometimes require.

▲ A GENUINE SHORTAGE

The one area in which a genuine shortage may soon be felt is in the Ph.D. ranks of science and engineering, particularly since university faculties are drawn from these ranks. Because of a shortsighted retirement policy forced through Congress by university administrators a number of years ago, colleges and universities are now beginning to feel the pinch.[15] But this too has a ready solution—in fact, two separate solutions. The first is simply to increase the support for doctoral candidates. About 70 percent of the full-time doctoral students in science and engineering are supported in one way or another, but generally at marginal levels when one considers that most are required to do some undergraduate teaching and grade papers as well. When a shortage threatens, one should expect market forces to apply here also. Adequate financial inducements for the best students to continue with their doctoral studies should come not only from the federal agencies that provide about 20 percent of such support today, but also from industry and the universities that would directly benefit from more doctoral candidates. If high-tech industry in the United States were to support only one additional doctoral candidate for every hundred million dollars of revenue, the number of Ph.D.s in the natural sciences and engineering could be increased by at least 25 percent. This would be a modest investment indeed for a substantial return over the long term, should industry become truly alarmed about an impending shortage of Ph.D.s.

But there is also a possible short-term remedy. Between 20 and 50 percent of all Ph.D. degrees in the natural sciences and engineering awarded in the United States, depending upon specific field, go to foreign nationals from countries like China, Taiwan, Korea, and Japan. Some of them are financed by their governments, but many are supported through U.S. research grants to universities. Foreign graduate students are found in all fields, but especially in technology and the sciences, where the American reputation for outstanding university research facilities is well known throughout the world. For the past twenty years or so the number of doctorates granted in the United States has remained roughly constant, but the fraction of foreign students earning these degrees has increased dramatically. For various personal reasons, but also because of U.S. immigration policies, roughly half the foreign students return to their homelands after graduation. It is well known that many of those who leave would prefer to remain in the States but are unable to secure resident status. While many Americans view the large number of foreign students in science and engineering as bordering on a national disgrace, they seem to forget that immigration in the past had provided much of the U.S. talent in these fields. Rather than discourage foreign students from studying here we should seek to encourage as many as possible of the best students to remain after completing their studies. It is shortsighted to export such talent to compete with us in the face of impending shortages of Ph.D.s. In fact, the suggestion has been made that foreign students who obtain graduate degrees here should be entitled automatically to immigrant status upon graduation—that, in effect, their diplomas should come with "Green Cards" attached.[16]

The chief complaint of those who deplore the huge influx of foreign graduate students in science and engineering is that the students who return to their homelands simply drain financial and academic resources away from American students, particularly minorities, who might otherwise seek careers in these fields. Some also assert that those students who leave after being trained here constitute a potential security risk to the United States. Such arguments might hold if adequate numbers of qualified American students were candidates for advanced degrees, but unfortunately this is not the case. The social forces driving foreign students to graduate work in the sciences today appear to be lacking in native Americans, as well as in many European nations, reflecting a trend toward careers in commerce, finance, law, and service industries, all of which are perceived as offering better prospects than science and engineering. A 1989 poll of undergraduates taken by the *Daily Telegraph* (London) revealed that careers in science and engineering are no longer considered prestigious, even among science and engineering students. In countries like Germany and Japan, where rigorous scientific training has been the norm for most students and science careers have always been highly prized, there now appears to be a trend away from careers in science and engineering toward business and financial services.

▲ A DESIRABLE SHORTAGE?

Most of the scientific and engineering communities accept the claim of impending Ph.D. shortages in their ranks and seem not to be greatly perturbed by it, perhaps thinking that shortages can only help raise their status and attract more students to the profession. But some, notably economist Paula Stephan, believe that a shortage may be helpful for quite another reason. Citing some interesting data, such as a huge increase in the relative number of doctoral scientists and engineers per million adults (over age twenty-two), from 320 in the pre–World War II year of 1940, to 2,000 currently, she questions whether their quality has been preserved, or whether a weeding-out process might actually be beneficial.[17] She argues that, as the number of doctorates grew, the proportion trained at the "best" schools naturally declined; and she concludes, based partly on comparisons of the numbers of publications early in their postdoctoral careers (specifically in particle physics), that "the younger scientific community we have today is not as productive as younger scientists 20 or 30 years ago"; she suggests that this decline in quality will likely continue for the next ten or fifteen years.[18] Another concern, says Stephan, is that the large number of Ph.D.s in nonprofit institutions competing for a disproportionate amount of grant support forces them to avoid the risks of frontier research and to focus instead on "safe" research areas. Finally, she points to the well-established fact that for the past two decades careers in science and engineering have failed to attract the very brightest students in our colleges, who more and more seemed to choose careers in business, law, and medicine.

Stephan concludes from all this that tightening the Ph.D. supply, rather than increasing it, would be better for the profession generally;

the demand for Ph.D.s would increase, as would salaries, of course, thereby attracting more of the best students into the field, and young Ph.D. scientists and engineers might feel more comfortable about taking risks in their choice of research problems.

Not all scientists and engineers would agree with Stephan's conclusions, but there is enough substance in her study to warrant serious thought. For one thing, the marked increase in the percentage of Ph.D.s in the general population since the 1940s could simply be the result of a like increase in demand brought on by the postwar surge in U.S. and free world industrialization. For another, equating Stephan's measure of productivity (number of publications per young science doctorate) to scientific quality is questionable, as is her assumption that creating a shortage would result in increasing the *absolute* number of top Ph.D. candidates. Also, her conclusion that tightening the supply of Ph.D.s would benefit the scientific and engineering professions generally, while it is possibly correct in the academic sense, fails to take into account the needs of industry or the welfare of society as a whole, where quantity must be considered as well as quality. Nevertheless, her thesis provides an interesting sidelight on the scientific manpower debate.

As should be evident from the foregoing, at least to the extent that a "crisis" in science education is predicated upon predictions of a future shortage of scientists and engineers, there is good reason to suspect its validity. What many fail to appreciate is that in the final analysis the size of the scientific work force is determined more by the external marketplace than by any internal initiatives of the science education community. This is not an educational as much as it is a social and commercial problem. Whatever tactics may be adopted by our schools to increase the supply of students in science and engineering, without a *clear* message from the demand side—that is, from future employers—to those who have a major hand in guiding students toward career objectives, that excellent career opportunities will be available, and that science and engineering continue to play prestigious roles in American society, such efforts are bound to fail. This means getting the message to the students themselves, to their parents, and to their career counselors,

Thus it appears that the "crisis" was born out of frustration by the science education community, frustration with the administration for its apparent lack of interest in science education and possibly with itself for want of a clear sense of direction following the collapse of the previous round of curriculum reforms. The mistake, however, was in seeking to correct the situation by setting such an unrealistic goal as scientific literacy without first being secure in its meaning.

There is a curious admixture in all this of the nation's need to provide a steady stream of scientists and engineers, and hence to counter any shortages in the supply of such specialists, whether real or perceived, and the educational politics surrounding the drive for general literacy in science. The need for science-trained professionals concerns only a small fraction of our high school population (variously estimated at 10 to 20 percent who claim to be interested in science or engineering careers), while the campaign for scientific literacy deals with *all* students. These numbers invite various rationalizations for general education in science for all students. They suggest, for example, that some 80 to 90 percent of the effort going into high school science education is little more than a shotgun approach to getting more students into the science pipeline. If this effort succeeded in developing some form of scientific literacy in our population, it would be understandable, but it fails this test as well, leading one to seriously question both the purpose of general education in science and the practice of it.

CHAPTER 2

Public Understanding of Science

The history of science is science itself; the history of the individual, the individual.
—Johann Wolfgang von Goethe (*Mineralogy and Geology*, 1814–1815)

▲   To appreciate fully the problems of science education, one must start from the premise that science is different from other disciplines—so different, in fact, that communicating it to the public at large poses some very special problems. It is not simply that the subject matter is different—that is obvious—but that to truly comprehend that subject matter requires, as we shall see, an uncommon mode of thought. Attempts to engage the public in the covenant of science are not new but date back to the earliest days of the new discipline, with results that were little more encouraging then than now.

▲   IN THE BEGINNING

When Plato discussed how best to educate the future leaders of society, he argued that to arrive at an idea of "the good," one must study, among other things, the hypotheses and basic concepts of the natural sciences.[1] His counsel turned to prophecy; for the next two dozen centuries, climaxing in the last four, when science truly prospered, Plato's words were to echo through government corridors and the halls of academe as scholars, statesmen, and educators debated endlessly on how to achieve this end. The debate still rages, focusing mainly on science education for the nonscientist, for the public at large, yet the goal of universal scientific literacy seems as remote as ever.

We learn from history, from mistakes of the past, and, because of its cumulative nature, this is particularly true of science. It must be a truism that we cannot teach what we do not understand; hence it is instructive to look at how the growth of science was inhibited in its early days by certain modes of thought that, while largely conquered by science, return to haunt us time and again in science education. Modern science, characterized by rational thought and by methods that in the main have led successfully to the understanding of natural phenomena, is relatively recent, having its origin in the seventeenth century. But its roots may be traced, by a sometimes tortuous path, to the ancient Greek culture of Plato and his followers. It is important here to distinguish between the orderly enterprise we know as science and the empirical, technological developments resulting from the human need to control our environment. Empirical technology, often termed "practical" or "applied" science, or simply "technology," and largely involving trial-and-error methods, dates back virtually to the dawn of civilization. However much these discoveries—for example, in metallurgy, ceramics, irrigation, and the mechanical arts—may have contributed to the growth of civilization, the methods and motives that led to them should not be confused with the structured enterprise by which we seek to confirm our modern views of nature.

It would be a mistake to conclude that the quest of understanding our environment dates back only three or four hundred years. Earlier civilizations were no less interested than our own in knowledge, nor less curious about the nature of things. The essential difference is that the earlier methods by which such knowledge was sought were inadequate to reveal the underlying causes of natural phenomena. The Golden Age of Greece, the fifth and fourth centuries B.C., saw the beginnings of a system of natural philosophy that was to dominate scientific thinking for centuries to come. This was the period of the famous teacher-pupil sequence: Socrates, Plato, and Aristotle—arguably three of the most remarkable individuals in the history of thought. All believed in the existence of universal or absolute truths, which the human mind could discover if only it pursued the proper methods. They looked upon knowledge not as a means to utilitarian ends, but as a means of satisfying human curiosity. And they held that *true knowledge*, as distinguished from knowledge derived via the senses, could be arrived at by purely formal methods, that is, by a system of logic. The use of some forms of logic appears to date back to the sixth and fifth centuries B.C., when philosophers first showed concerns over the internal consistency of their arguments. But it was Socrates and his student Plato, who put deductive logic to such skillful use, and Aristotle, the best known of Plato's pupils, who perfected the classic logical device known as the "syllogism."

The Socratic method of reasoning found ready acceptance among the knowledge-loving Greeks, who became masters of deduction—and the victims of its weaknesses. We all use deduction in our daily affairs. It is the process by which one proceeds logically to the solution of particular problems from premises assumed to be true. Thus, in its simplest form, if one asserts that a is greater than b, and b greater than c, it follows that a is greater than c. The conclusion is a *logically* necessary consequence of the premises, and while the deduction is clearly correct, the *truth* of the conclusion rests upon the reliability of the initial premises. It should be evident that deductive reasoning does not necessarily lead to new basic knowledge, for it provides no recipe for testing the truth of the basic assertions, for example, that a is indeed greater than b, or b greater than c. This is a major failing of modern science education for the general public: it portrays deductive reasoning as the cornerstone of science, when in reality it is only a stepping-stone in the process of reaching reliable conclusions about nature.

The syllogism, which all schoolchildren learn at some point, represents a particular form of deductive argument having a structure frequently found in ordinary discourse. As in the symbolic example given above, it consists of a general premise, assumed to be valid, followed by a statement that applies the premise to a particular case, and finally a *logical* conclusion. As an example, we might construct the following syllogism to arrive at a typical Aristotelian explanation for the acceleration of falling bodies:

The traveler hastens as he approaches his destination.
A falling object may be likened to a traveler. . . .
Hence, falling objects speed up (accelerate) as they approach the Earth.

The argument is obviously defective in its premises. Even granting the (questionable) validity of the initial premise, the analogy drawn in the second has no basis in fact and ascribes human qualities to the object. Yet the conclusion follows logically from these statements, and the hypothetical example illustrates how Aristotle's disciples would employ his deductive system to "account" for natural phenomena. Such reasoning from false premises to *correct* conclusions is unfortunately all too common, even today, as we shall see later, in the area of explanation. The conclusion is correct, of course, because it stems from observation, and the premises are designed as a logical explanation for the conclusion, without regard to their validity.

The example given represents but one of many variations of the syllogistic form of argument, in this case being more an example of *dialectic* than of *scientific* discourse. In the final analysis the latter seeks to reason from true premises while the former only from *probable* or even *plausible* statements. The formal logical methods by which such reasoning may be examined for self-consistency were already developed to a high degree by the start of the Hellenistic period (with the death in 323 B.C. of Alexander the Great, who had been tutored by Aristotle), during which they formed the foundation for the remarkable mathematical proofs of Euclid (ca. 323–ca. 285 B.C.) and Archimedes (287?–212 B.C.). But however useful the deductive method may be in pure mathematics, it cannot alone serve as a means of understanding nature. The essence of any natural science is to account for nature in the simplest possible terms, that is, to try to reduce all that we observe to basic principles or causes. This is what we mean by *explanation* in science and this is the way we discover new scientific knowledge. But how does science seek out the basic causes of what we observe? How do we establish the *truths* of our initial premises when reasoning deductively? Here, the Aristotelian procedure fails and one must look to other methods.

Aristotle differed from Plato and Socrates in part by his marked interest in natural phenomena and his higher regard for practical matters. But he was not a great scientist (in the modern sense), and it will be instructive for our purposes to see why his brilliant intellect failed him in science. Unlike most of his contemporaries, who believed that the secrets of nature could be discovered through pure thought, Aristotle was not adverse to experimentation, but he could hardly be considered a thorough experimenter. His chief qualifications for being regarded as a "scientist" rest on his careful, systematic observations in descriptive natural history. But regarding the nature of the physical world, Aristotle held the most naive and confused views (by present standards). Much as he contributed to the early development of biology, it is because of his flawed physical reasoning that one regrets his very strong influence over succeeding centuries of scientific thought. There is little doubt that it was largely his authority that delayed so long the evolution of such areas as dynamics, atomism, and astronomy. Some two thousand years later, we find the founders of modern science, such as Gilbert, Copernicus, Galileo, Boyle, and Newton, having to reject the prevailing doctrines of Aristotle before setting science on a firm foundation.

There are several possible causes for Aristotle's unsound doctrines in physical science, but two stand out—especially because they remain important factors in the success (or failure) of modern science education. First, he failed to make effective use of the process of *induction*, which is the inverse of the deductive method and consists essentially in proceeding from the particular to the general—in reasoning from a limited number of observations to a general conclusion embracing all similar events. It is by this method that modern science establishes the initial premises (postulates, theories, etc.), from which one can then proceed, by deduction, to account for particular experiences, thereby testing the reliability of the induction itself. Aristotle was familiar with the inductive method; in fact, he was the first to outline its principles. Perhaps he distrusted it as a means of gaining knowledge, or perhaps he simply failed to recognize its place in scientific reasoning. At any rate, he did not make use of this highly significant form of argument. Instead, the generalizations that were to form the initial premises in his deductive arguments were arrived at either by pure intuition or by appealing to the ends or purposes they served (generally human), that is, from teleological arguments. This is not to say that intuition has no place in inductive reasoning; but it is intuition born of experience rather than the introspective inventions of Aristotle that distinguishes the modern use of induction in science. As Bronowski put it, "Inductive reasoning . . . extends the experience of the past into the future."[2]

At this point a word is in order on the use of teleology as a mode of explanation, especially as it is employed so often in the biological sciences. Teleology consists of explanation in terms of end purposes or functions, as opposed to the initial causes one invokes in the physical sciences. One is often tempted to distinguish the two by asserting that teleology explains the present in terms of the future, while a true scientific explanation accounts for the present in terms of the past. This is misleading, however, for while many scientists (including biologists) believe that in the end it will prove possible to reduce all of biology to explanation in physico-chemical terms, because of the complexity of even the simplest of organisms, teleological explanations are nevertheless useful in teaching about them. Thus, in explaining various animal organs to students, it is customary to cite their function or purpose; for example, in times of stress the brain needs more oxygen, so the heart rate increases, or the function of chlorophyll in green plants is to enable photosynthesis to occur. Still more common is the use of teleology in evolutionary explanations, where evolution is often confused with adaptation, for example, that humans developed hands to better grasp tools, or sea animals developed limbs so they could navigate on land.

Such goal-directed explanations, while not reaching down to root causes, can be very helpful at the student level, although without careful qualification they can lead students to a gross misunderstanding of the nature of scientific explanation.

While the physical and biological sciences continue to move closer together in terms of explanation, they do not yet coincide; hence it should not be surprising to find them using different modes of explanation. Aristotle saw no difference between the two as to type of reasoning—thus his explanation of physical phenomena in terms of teleological or end-causes, which no doubt accounts for his failures in the physical sciences.

The second reason for Aristotle's failure to produce any enduring scientific principles is that his physical arguments were largely nonmathematical; they were qualitative rather than quantitative and lacked the abstraction that became the power of modern science. Nor did he design critical experiments to test the conclusions deduced from his premises, which, we shall find, has become the central feature of science. Leonardo da Vinci (1452–1519), the remarkable Florentine painter who was among the first to hammer at the crumbling walls of medieval science, held that true science began with observation; if mathematical reasoning were applied, greater certainty might be reached, but "those sciences are vain and full of errors which are not born from experiment, the mother of all certainty, and which do not end with one clear experiment."[3]

Practicing the science of one's day, such as it was, was one thing, but how should this knowledge be communicated to the public? Cicero, the famous Roman statesman and orator, apparently much impressed by Aristotle's writings, boasted that the ideal orator, of which he clearly considered himself to be the prime example, could explain anything, including Greek science (of which he was largely ignorant), to anyone in ordinary Latin terms, often better than could one who made that subject his life's work.[4] Indeed, Cicero probably did an admirable job of explaining the few scientific concepts of which he was knowledgeable, mainly because the nature of science in his day lent itself so well to speculation and rhetoric. Moreover, Cicero, and Plato before him, would not be greatly concerned with the masses but with the small circle of scholars who shared their thirst for knowledge and had the means to pursue it.

The fact is that throughout the Hellenistic period and the Middle Ages, as science grew and finally as it blossomed in the scientific revolution of the Renaissance, all efforts at educating the public in science were aimed at the small segment of the general population that was at least literate, if not formally educated.[5] Even with a formal education the science that was taught was at best rudimentary. When Galileo wrote his masterful *Dialogue* in 1630, which brought him before the Inquisition, he was trying to convince the common citizen of his views.[6] He wrote in Italian rather than the more scholarly Latin, and in a style distinguished by its readability and literary elegance. It was a style quickly outmoded by the more formal scientific writing to which we are now accustomed, perhaps unfortunately so for the cause of scientific literacy. Early in the seventeenth century, at the start of the scientific revolution, a contemporary of Galileo's, the great English Renaissance philosopher Sir Francis Bacon, deploring the fact that the sciences then available were useless for making practical discoveries, asserted that the real purpose of science was to "improve man's lot" rather than to provide "pleasure of the mind."[7] Bacon could not have been more prophetic; he foresaw what most nonscientists would ultimately find of value in science—its utility. Indeed, the argument as to which kind of science is more appropriate today for developing literacy in the general public—pure or applied, theoretical or practical, in effect, science or technology—still prevails in the science education forum.

In Bacon's day intellectual life was dominated by a small number of scholars and clerics, theorists rather than doers, individuals who perpetuated their knowledge by the rule of authority, mainly by superstition, religious dogma, and the philosophical tradition of Plato and Aristotle. Bacon was unimpressed by his peers and rebelled against what he regarded as a sterile system of scientific thought, one that did not rely on experiment and hence could not provide the foundation for new inventions and new industries that might benefit all citizens. He also believed it important to diffuse scientific knowledge to the masses, which some interpreted as a cleverly conceived attack on the authority of revealed religion. His skepticism foreshadowed the Enlightenment a century later, when a number of critics railed against superstitious beliefs, irrationalism, and the sometimes stifling influence of church doctrine on the pursuit of knowledge.

▲ THE SCIENTIFIC/INDUSTRIAL REVOLUTIONS

With the scientific and industrial revolutions in full swing, the eighteenth and nineteenth centuries saw greatly increased efforts aimed at popularizing science. Voltaire, probably the most famous Enlightenment skeptic, translated Newton for his French readers, and also performed for his friends some of Newton's spectral demonstrations using the light from the Sun. The electrical experiments of Galvani and Volta, and later of Faraday, became favorite subjects in the royal courts of Europe, and even Napoleon I, it is said, showed an interest in them. This is not surprising, as the Napoleonic reforms of public education made science a permanent part of the basic curriculum and had a worldwide effect on educational theory. In fact, Napoleon founded the world's first technical school in Paris in 1793 (the famous École Polytechnique). It should be noted, however, that Napoleon's interest in science education was wholly pragmatic; he looked upon it as a means of better preparing his armed forces for the battlefield, and only those students who chose military careers could receive the full benefit of the science education that was then available in the lycées.[8]

Napoleon was not alone in this Baconian view of science. By the early nineteenth century it was obvious to most if not to all politicians and statesmen that science was more than an interesting curiosity; it could play an important practical role, not only in war but in the economy of nations. Hence support of science and its inclusion in the curricula of secondary schools and colleges began taking hold all over western Europe and North America. As Thomas Huxley pointed out in 1868: "The politicians tell us, 'you must educate the masses because they are going to be masters.'"[9] The general public seemed to exhibit more than a casual interest in science through their reading of books and their attendance at public lectures, and through the science "shows" (mostly electrical) that were popular at the time, both in Europe and America. Among these were the famous Christmas lectures presented by the Royal Institution in London, perhaps the best known being Michael Faraday's course of lectures on "The Chemical History of a Candle," delivered before a predominately juvenile audience in 1859–60. Was this seemingly popular interest in science any more than curiosity about something new and mysterious? Probably not, but it must be realized that those who could take advantage of such offerings comprised only a small and privileged fraction of the general population, namely, individuals having some level of education and stature in society. It cannot be concluded that the public at large was taken with science; most, in fact, if they thought about it at all, regarded science as a form of black magic—or even worse, witchcraft.

Nonetheless, as measured by some norms, public interest in science during the nineteenth century might appear to have been greater than it is today. This is not true, of course, except that the novelty has worn off and the public is much more casual about science. The products of science, mostly absent in the last century, are commonplace today. We no longer see many public science demonstrations in lecture halls and theaters, but for those interested enough to take full advantage of it, television and other mass communication media offer much greater opportunity for public understanding of science today than at any time in the past. Yet despite the enormous impact of the products of science on everyone's daily life, and the greatly increased opportunities for the public to learn about science, the response of the general public has been generally disappointing; the number of Americans taking advantage of such opportunities is relatively small. It is in this sense that one might argue the proposition that, given the same opportunities for access to science in the nineteenth century, the public response would have been proportionately far greater. Probably true, but such speculation has little value except as an indicator of the changes that have occurred both in science and in society. These changes, we will find, make science, or rather the attempt to understand science, much less attractive to the general public of today.

▲ SCIENTIFIC LITERACY: SOME EARLY VIEWS

At least some of the public interest in science in the nineteenth century was triggered by the writings of the famous English philosopher, educator, and ethicist Herbert Spencer, and the controversy surrounding his views. The best known of Spencer's essays on education, "What Knowledge Is of Most Worth," published in 1855, clearly demonstrates his strong personal bias for science. Yet here was a classic example of logic gone awry. Spencer reasoned, quite correctly, that the purpose of education was to prepare one for a complete life. He then went on to analyze the activities and needs of everyday life and how the various forms of study contributed to these, finally concluding that of all subjects a knowledge of science was the most useful preparation for living. Spencer summed up his views on the role of science in education in the following words:

> Thus to the question we set out with—What knowledge is of most worth?—the uniform reply is—Science. This is the verdict on all counts. For direct self-preservation, or the maintenance of life and health, the all-important knowledge is—Science. For that

indirect self-preservation which we call gaining a livelihood, the knowledge of greatest value is—Science. For the true discharge of parental functions, the proper guidance is to be found only in— Science. For the interpretation of national life, past and present, without which the citizen cannot rightly regulate his conduct, the indispensable key is—Science. Alike for the most perfect production and highest enjoyment of art in all its forms, the needful preparation is still—Science. And for purposes of discipline—intellectual, moral, religious—the most efficient study is, once more —Science.[10]

Spencer went on to deplore what he saw as the lowly estate of science in the contemporary education of his day:

And yet this study, immensely transcending all others in importance, is that which, in an age of boasted education, receives the least attention. While what we call civilization could never have arisen had it not been for science, science forms scarcely an appreciable element in our so-called civilized training. Though to the progress of science we owe it, that millions find support where once there was food only for thousands; yet of these millions but a few thousand pay any respect to that which has made their existence possible.

Spencer's complaint about the status of science in the educational curriculum of his day is often echoed by science educators today, but few would dare suggest that an education based *primarily* in science could be the answer to all the world's problems. Indeed, it is hard to understand how Spencer, audacious and in some ways scientifically naïve though he was, could have espoused such a bizarre point of view. He was, after all, perhaps the most-discussed English thinker of the day. He did not fashion his thesis out of thin air, however. Others before him had warned that lacking adequate support from the privileged class, British industry was in danger of falling behind its competitors, leading Spencer to conclude that only by radically altering Britain's traditional educational structure could its world dominance be preserved.

Needless to say, his essay evoked a violent response. Later, in his autobiography, Spencer recalls that "when this essay was written, its leading thesis, that the teaching of the classics should give place to the teaching of science, was regarded by nine out of ten cultivated people as simply monstrous."[11] Also needless to say, it failed to achieve his hoped-for effect of displacing the classics in England's educational system. We shall see later that the current thinking of some prominent educators in England essentially parallels that of Spencer's regarding the role of science in sustaining British industry, at least at the pre-university level, resulting in a recent reform of its national curriculum to position science as its centerpiece.

Spencer was not alone in advocating a dominant role for science in education. Other strong voices followed, notable among them being Thomas Huxley, the well-known English biologist and educator and the foremost advocate at the time of Darwin's theory of evolution. Huxley apparently had little respect for Spencer's scientific work, but with Spencer he was equally outspoken in favoring science over the classics in the school curriculum, except that his rationale was somewhat more moderate. In an 1880 address entitled "Science and Culture," Huxley offered this opinion:[12]

> For I hold very strongly by two convictions—the first is, that neither the discipline nor the subject matter of classical education is of such direct value to the student of physical science as to justify the expenditure of valuable time upon either; and the second is, that for the purpose of attaining real culture, an exclusively scientific education is at least as effectual as an exclusively literary education.

This was a quarter century after the argumentative Spencer had so infuriated the Victorian literati, from which it can be seen that the "two cultures" controversy, made famous almost a century later by Lord Snow, was already blossoming. That Huxley was mindful of this conflict is clear from his further remark:

> I need hardly point out to you that these opinions, especially the latter, are diametrically opposed to those of the great majority of educated Englishmen, influenced as they are by school and university traditions. In their belief, culture is obtainable only by a liberal education; and a liberal education is synonymous, not merely with education and instruction in literature, but in one particular form of literature, namely, that of Greek and Roman antiquity. They hold that the man who has learned Latin and Greek, however little, is educated; while he who is versed in other branches of knowledge, however deeply, is a more or less respected

specialist, not admissible into the cultured caste. The stamp of the educated man, the University degree, is not for him.[13]

For Huxley to suggest not only that in a cultural sense a scientific education was *at least* the equal of a classical education, but that for a student of science it was a waste of time to study the classics, was sheer academic heresy in his day. Today, while many science students might sympathize with Huxley's view on spending one's time in classical studies, or, for that matter, on any studies that take them away from science, few academic scientists in liberal arts or engineering faculties would agree with it, at least not publicly. We shall return to the "two cultures" argument later, as it bears significantly on the basic issue of scientific literacy.

Huxley's vision of science was only partly intellectual. He spoke of science as something worth knowing simply because it was gratifying to the intellect; as such, it was a viable substitute, in his view, for an exclusively literary (classical) education, which he, among others, regarded as an arrogant monopoly. But he also seemed to share Spencer's conviction, albeit with somewhat less naïveté, that the scientific revolution would soon cause a kind of secular revelation that would so alter society's view of revealed truth as to make science the dominant factor in the future affairs of man. Like Spencer, Huxley's views on an exclusively scientific education did not go unchallenged, notably by the English poet Matthew Arnold, who was Inspector of Schools at the time. In his Rede lecture of 1882, Arnold argued not against making science a part of the school curriculum, but for the need of the nonscientist to be acquainted only with the factual results of science, not with its methods.[14] Not surprisingly, Arnold's position found much greater support than did Huxley's at the time; it would not be well received by educators today, certainly not by science educators and their science mentors. Yet the truth is that, if anything, today's ordinary citizen is more knowledgeable about scientific facts than about its methods.

In fact, there is growing evidence that even the U.S. Congress, science funding agencies like the National Science Foundation, and industries that traditionally have supported basic science in the past are now demanding more immediate practical results from investments in science research. The newest catch words are "focused" and "targeted" research, generally taken to mean technology. And the race is on in all industrialized nations, no longer as much for the prestige of Nobel and other such prizes as for technological supremacy. Much as the scientific community deplores this trend, knowing that without basic research the well will eventually go dry, it must learn to accommodate the technological demands of society. At least one good may come of it: science and the public may at last begin speaking a common language.

There were many others who shared Huxley's nonutilitarian view of science then, as do many academic scientists today. Their objectives were clear: in an age when virtually everyone looked to the practical benefits of science, for science to share the educational spotlight with the classics it would have to be shown to offer at least similar aesthetic and intellectual values. J.B.S. Haldane, another noted English biologist and educator, said some fifty years later: "Science is vastly more stimulating to the imagination than are the classics."[15] And the noted French mathematician and philosopher of science Henri Poincaré, distinguished between the beauty one can find in the scientific study of nature, and the aesthetic beauty found in nature itself—or in the creative arts generally: "Of course I do not here speak of the beauty that strikes the senses, the beauty of qualities and appearances; Not that I undervalue such beauty, far from it, but it has nothing to do with science; I mean that profounder beauty which comes from the harmonious order of parts and which a pure intelligence can grasp."[16]

In the century that followed, this lofty theme was to be repeated time and again by prominent scientists and educators, seeking, with Huxley, to portray science as an essential part of a liberal education. But where such words would strike resounding chords in the minds of scientists, they usually fell upon deaf ears in the general public.

▲ SCIENCE AND GENERAL EDUCATION IN EARLY AMERICA

Serious efforts toward achieving public understanding of science in the United States had to wait until science instruction became an integral part of the secondary school and college curricula. A number of prominent voices spoke out for education in science early in the founding of this nation. Ardent champions then were Benjamin Franklin and Thomas Jefferson, who, with Bacon, saw science as important mainly for its utilitarian values in a developing country. In fact, science instruction in the United States had its beginnings in the Philadelphia Academy of Science founded by Franklin in 1751,[17] and Jefferson, while a member of the Virginia legislature in 1779, introduced a bill proposing science instruction throughout the elementary, college, and university grades.[18] Franklin's view of science, in common with that of most other intellectuals of his time, was wholly pragmatic; the aims of science instruction were to be descriptive, utilitarian and religious, and the mode of instruction was generally catechismal. He rationalized science education for farmers that they might produce better crops; for would-be merchants so they might better understand the commodities they would sell; for intended craftsmen so that they might learn the use of new materials; and for those going into the ministry so that they should better understand proofs of the existence of God.[19] Yet strangely enough, for all his insistence upon the practical aspects of science in education, Franklin had a strong analytical bent, as evidenced by his theoretical contributions to the basic understanding of electricity.

That nineteenth-century science in America was mainly utilitarian was also noted by the French writer Alexis de Tocqueville, when he visited in 1831. In his *Democracy in America* he wrote: "In America the purely practical part of science is admirably understood, and careful attention is paid to that theoretical portion which is immediately requisite to application. On this head Americans always display a clear, free, original and inventive power of mind. But hardly anyone in the United States devotes himself to the theoretical and abstract portion of human knowledge."[20]

It should not be surprising that in eighteenth- and nineteenth-century America, when the new nation was struggling to develop its own land and its own industrial and economic strengths, its science and science education were focused on the practical rather than the abstract; indeed, it would have been folly to do otherwise. The examples given above of pre–nineteenth-century interest in science and science education in America were exceptions rather than the rule. Science was not a prominent activity in America at the opening of the nineteenth century; the eighteenth-century enthusiasm for science among the educated upper class of the American Revolutionary era appears to have waned. The number of full-time practicing scientists was pitifully small; Burnham places this number at only twenty-one individuals who earned their livelihood as scientists in 1802.[21] As further evidence of the sorry state of American science at the time, Greene notes that of more than thirteen hundred books printed in the United States and listed in a bookseller's catalog in 1804, "not more than twenty can be considered works of science," and most of these were textbooks.[22] But this changed soon afterward, as the practicing scientists began setting their specialized interests apart from those of well-educated intellectuals generally. An important measure of a nation's interest in science is the number and variety of professional societies it promulgates, and several of these were established in America early in the nineteenth century. The Columbian Chemical Society of Philadelphia was founded in 1811, the Academy of Natural Sciences of Philadelphia in 1812, the Lyceum of Natural History in the City of New York (later to become The New York Academy of Sciences) in 1817, the American Institute of the City of New York in 1828, and the Boston Society of Natural History in 1830. It was not until 1843, however, that the first technical university, Rensselaer Polytechnic Institute (RPI), was founded in America, in Troy, New York, with several others following during the next few decades.

Another measure of the prominence of science in the eighteenth and nineteenth centuries is the growth in numbers of scientific journals, first in Europe and then in the United States. The first two scientific journals were founded in 1665, one in France (*Journal des scavans*), and the other, the well-known *Philosophical Transactions,* in England. By 1700 about thirty scientific and medical journals had been established, all of a general scientific or medical nature, but by the start of the nineteenth century specialization set in and the publication explosion took off. At the turn of the twentieth century there were about two thousand scientific journals published regularly worldwide; this number grew to about ten thousand by the start of World War II, and may be estimated from published growth rates at some seventy thousand today.[23] Of these, it appears that about 10 percent are published in the United States.

It was clearly a period of ferment in American science—one that seemed to match the general feeling of strength and accomplishment of the American people following the War of 1812 and the growth of political democracy. To most people in the United States at that time, science was a curiosity, as it was in Europe, and some competent scientists, as well as quacks, played on the layman's interest in the mysterious and unfamiliar. Burnham gives an excellent account of the institutions that sprang up to popularize science in this period, noting that the public lecture became the principal means of conveying science to at least a part of the public.[24] These ranged from science "shows" focused on the spectacular to serious attempts to inform the public on developments in science through single lectures or the equivalent of full courses spread over a period of weeks. Burnham cites the advice of one such lecturer, Amos Eaton, to his fellow science lecturers who would enter upon such a venture. Eaton was a successful chemistry lecturer in New England and New York around 1820, and apparently felt it necessary to preserve the integrity of his calling:

> In commencing an itinerating course, let the Clergymen, Doctors, Lawyers, and other principal men in the village or district, send printed cards (prepared by the lecturer) inviting the citizens to attend a gratuitous lecture. At the first lecture . . . the plan and object of the proposed course should be illustrated by striking experiments. But never introduce those blazing, puppet-show-like experiments, common with quacks and imposters. . . . Itinerating lectures on Chemistry, Natural Philosophy (including astronomy), Geology and Botany, are of great use to small villages and country districts where permanent courses cannot be supported.[25]

These itinerant lecturers charged for their appearances, of course, and their audiences were mainly the educated citizens of means, but what of the masses, most of whom had very little formal education? Some were no doubt sufficiently interested to attend such courses, but most had to get their knowledge of science from the newspapers and almanacs that had already become popular in the eighteenth century.

The practical goals of science instruction as set forth by Franklin and his colleagues persisted until the late nineteenth century (about the 1880s), when a marked change in objectives emerged. As George Daniels has pointed out, American science reached a peak following the Civil War, the first conflict to bring science and technology so sharply to the government's attention.[26] The number of colleges and universities had by then increased to about four hundred, each providing some instruction in science as well as having staffs and facilities that could be used in support of the war effort. So important had science become, in fact, that both state and federal agencies were spending fairly large sums on scientific research, and there was even serious talk about establishing a cabinet-level Department of Science. This was never accomplished, of course, but periodically since then the issue has been revived in various forms. After World War II, the position of scientific advisor to the president was established by President Truman in 1951, and in 1957 President Eisenhower created the President's Science Advisory Committee (PSAC), with the chair of this committee to serve as the president's special assistant for science and technology. Finally, in 1976, by act of Congress, the Office of Science and Technology Policy was established, with a director who, while not a member of the White House staff, would also serve as science advisor to the president. Soon after the war's end the responsibility for supporting research and education in science and engineering was largely given over to the newly formed National Science Foundation, although the U.S. Office of Education (which, in 1979, became the Department of Education) continued to support activities in science education as well, mainly through assistance to state and local educational agencies.

Because the fortunes of science education are so closely linked to those of science itself, it is interesting to speculate on what might have been the effect on both had a cabinet-level department been established following the Civil War. Clearly, with the attention that would have been accorded science by such a high-level federal agency, one should expect it to take a different turn. Had this occurred, it is almost certain that the state of science education today—and possibly of scientific literacy—would be different than it is. The values of a society are often most clearly reflected in its educational structure, that is, in the provisions it makes for training its future citizens. Thus, had science enjoyed cabinet status in the U.S. for the past century, one can easily imagine the ultimate effect this would have had on science education. It is probably not too extreme to theorize that we would now have many more scientists and engineers, for with the incentive provided by government support our technology-based industry would undoubtedly have grown more quickly, and a more sophisticated science education infrastructure to support this emphasis on science, possibly even including a system of national science high schools and colleges, might have evolved. Our basic overall scientific enterprise is presently second to none, but had there been a Department of Science in the federal government, some fields of science (and the technology and educational structures that derive from them) might well be a generation or more farther advanced at this point in time.

The late nineteenth century also saw science developing as an independent intellectual force, breaking away from both the utilitarian philosophy and the natural theology that had fettered it since the Copernican revolution. It was no longer necessary to justify one's pursuit of science in such terms as seeking a solution to some practical problem or illustrating some biblical text. Instead, science could now be practiced for its purely intellectual value or as a contribution to new knowledge—pure (theoretical) science, rather than applied science, the latter having come to imply (unreasonably so) a lower form of intellectual pursuit. With this new view of science one would expect some changes in the goals of science education, as indeed there were, at least in theory if not in practice.

The major change in educational philosophy that occurred in the 1880s was the giving way of religious and utilitarian aims to education designed for training of the mind. Not that the factual or practical goals of science were abandoned; far from it, but as pointed out in the 1938 Report of the Committee on the Function of Science in General Education, in referring to the late nineteenth century, "this was the heyday of faculty psychology and the doctrine of formal discipline,"[27] which held that the study of science offered particular merits not found in other studies. These included developing the student's mental discipline through training the faculty of observation, promoting one's concentration of thought and energy, and training the senses through the manipulation of materials, that is, hands-on laboratory experiences.[28] Factual information, memorization, abstract reasoning, problem-solving, but usually not real-life problems—all were characteristic of science education in this period. Indeed, while the principal educational theories associated with the period were displaced only thirty years later (ca. 1910), one can find many of these same characteristics, however much they may be deplored, in contemporary science education. The study of science came to be regarded primarily as preparation for college in this period, but the Committee of Ten saw no reason to provide different courses for those planning to go on to college and the bulk of the students, whose formal education would end with high school. This followed from the premise that the mental discipline provided by secondary education should prepare one for a vocational life as well as for future education, a premise that would be endorsed by many educators today. The Committee of Ten, which was established by the National Education Association in 1892 to study and possibly recommend some degree of uniformity in the admission requirements of colleges, understandably had a considerable impact on the high school curriculum, even though, unlike many science educators two decades later, the Committee believed that the *primary* function of secondary schools was not to prepare students for admission to college. Its report, as might have been expected, greatly influenced the design of high school science courses but did little to clarify the goals of science education. Since then there have been numberless committees and hundreds of reports on the purpose and perceived problems of high school science education, yet we seem little more advanced in our understanding of why we teach what science to which students.

The rebellion against a purely classical (or literary) education, triggered by Spencer in mid-nineteenth-century England, crossed the Atlantic early in the twentieth century, although in a more moderate fashion. Instead of calling for the replacement of the traditional curriculum by one based in science, the strategy was simply to add science to the curriculum, initially as an elective subject but ultimately as part of the compulsory core curriculum in high schools and colleges. However, the two threads of science education, the "why" and the "what," soon became intertwined and have remained so to this day. *Why* should science be required of all students and *what* should it consist of? In early America, we have seen, the answers were easy; science education should be practical and utilitarian. But as the precollege science curriculum came under the influence of university scientists, its theoretical aspects became emphasized more and more and we now have no formal clear-cut guidelines.

▲ SCIENCE EDUCATION IN THE TWENTIETH CENTURY

Science did not become a major factor in general education in the United States until well into the twentieth century, for the simple reason that very few students continued their education beyond the elementary level. At the turn of the century only a small percentage of high school age youths actually attended school; by midcentury, however, practically all youngsters of this age were enrolled in high schools. And the U.S. scientific enterprise continues to grow, as it does in all highly industrialized nations. In 1956 there were fewer than a million scientists and engineers in the United States; today this total is variously estimated at three to four million. Early in the century, the expressed purpose of high school science alternated between general education and preparation for college, and between satisfying the practical needs of students and the intellectual demands of society; now it is once again considered a part of the general education of all students, except that the current goal, scientific literacy, is no better defined than were the earlier goals of science education for the general (nonscience) student. However, there appears to be rather general agreement among science educators on the importance of relating school science to the everyday experiences of students. We have yet to see whether in the long run these approaches will produce any better results in terms of scientific literacy than previous ones. But despite the continued growth in U.S. science activity, the prospects of achieving a common literacy in science do not appear very bright.

We have already characterized the decade of the 80's as one of particular turmoil in science education. The early part of the decade saw a complete letdown from the period following World War II, when science education reached new heights, both in government support and curriculum reform activity. We have also noted that the temporary collapse of the science education movement was brought on by the confluence of two factors. One was the sense that perhaps the new curricula developed in the previous three decades had missed their mark, for student outcomes seemed not to have changed much. The other was the decision of the first Reagan administration to dismantle the Science Education Directorate of the NSF, which had been the major support of curriculum reform.

Except for clusters of individuals who constantly deplore the sad state of functional literacy generally in the United States, and some groups (scientists, science educators, industrialists, and a few government agencies) who point to the obvious fact that scientific literacy is in a very bad way, and the executive branch, which now seeks to improve education generally by providing advice and moral support to the states and local communities, but not enough in the way of financial support, there appears to be no concerted movement in the United States to resolve either social deficit. There is a great deal of lip service available but little evidence of a meaningful drive toward solutions. In an editorial published at the end of the 1980s, Mortimer Zuckerman commented on the magnitude and import of the problem:

> In 1889, a person was judged literate if he could sign his name—enough for the farm-and-buggy economy. In the machine economy of 1939, it meant completing the sixth grade. Today, the Information Age of computers and high technology requires a bare minimum of reading and writing skills at the high-school-graduate level. Changes in workplace needs are so dramatic and unpredictable that people must be ready to adapt to jobs that did not even exist when they were in school.
>
> There are 25 million Americans who cannot read or write at all. An additional 45 million are functionally illiterate—without the reading and writing skills to find work—and that number is growing by more than 2 million a year.[29]

If indeed one-third of our total population is functionally illiterate or worse, and, as we shall see, well over 90 percent is scientifically illiterate, why is there so little *organized* national clamor over this state of affairs? The most obvious answer is that the problem has yet to be acknowledged officially and materially at the highest levels of government, and this is unlikely to occur until most citizens are willing to accept the fact that we have some serious deficiencies in our social and educational goals.

CHAPTER 3

The Nature of Science

When you put your hand in a flowing stream, you touch the last of what has gone before and the first of what is yet to come.
—Leonardo da Vinci (*The Notebooks*, 1508–1518)

▲   To gain some insight into the problem of achieving a common literacy in science, we must first examine the nature of science itself. The reason, as we alluded to earlier, and will examine more fully in this chapter, is that science differs so from all other intellectual activities as to require a mode of thought that is out of keeping with one's everyday experiences. This point will be emphasized repeatedly: our failure to develop a scientifically literate public boils down to the fact that most students (and adults), while they may find science interesting, also find it difficult and unrewarding to learn. An obvious conclusion is that we have been aiming too high. Contrary to what most science educators contend, knowing science in the formal academic sense may *not* be a necessary condition to attaining scientific literacy in the social sense. However, knowing what science is about *is* prerequisite to such literacy. The distinction may seem subtle at this point, but is nonetheless important. We will never get the mass of our population to understand science in detail, but we may be able to instill some understanding of how the enterprise works and how scientists practice their discipline— enough, one hopes, to serve the societal purpose of scientific literacy.

▲   NO SIMPLE DEFINITIONS

Civilization has always prized bold ideas, whether in the arts, literature, politics—or in science. Throughout history great ideas stand out as focal points for new systems of philosophy, new religions, new modes of thought, even new societies. Their counterparts in the sciences play a similar role. The big ideas in science are civilization's response to the challenge of nature, our way of trying to account for familiar observations in terms of a relatively few basic schemes or "rules" that help to unify broad ranges of experience.

Science is our formal contact with nature, our window on the universe, so to speak. It is a very *special* way that humans have devised for looking at ordinary things and trying to understand them. Whatever the motive, whether simple curiosity, which it pretty much was at the beginning, and still is in many cases, or our ultimate desire to modify or control nature, science boils down in the end to asking the proper questions of nature and making sense of the answers. Why a special way? Because experience has shown that our everyday modes of inquiry are inadequate to reveal the underlying causes of natural phenomena. Science is not simply a matter of accurate and detailed descriptions of things, or of extending our senses through the use of scientific instruments, both of which seem to be how most people look upon it. These are merely steps—important ones but nevertheless only a means to a much larger objective: the design of conceptual schemes, models, and theories that serve to account for major segments of our experience with nature, and ultimately form the bases for all explanation in science. The handful of major conceptual schemes that constitute the pinnacle of explanation in science must be classed among the greatest of civilization's intellectual achievements. While they may differ greatly in subject matter within the broad field of science, these conceptual schemes have one thing in common: they are not susceptible of *direct* experimental verification. Thus the assumption that matter is composed of small discrete particles (atoms or molecules), which is basic to the kinetic-molecular theory of matter that all students are exposed to at some point, is not subject to proof of the kind that might be considered "direct." Or, judging by the controversy that still surrounds the theory of evolution in some quarters, clearly the evidence for it is not considered sufficiently convincing by everyone, although it is widely accepted by the scientific community.

The same is true of all major conceptual schemes; to scientists they are essentially "articles of faith." Our confidence in them rests upon the degree to which they help us to account for our experiences with nature in an intellectually satisfying fashion; and the wider their range of application, the stronger becomes our belief in their validity. This is not to say that conceptual schemes are infallible; they are subject to almost the same uncertainties as any other conceptual ideas. But those that persist after being subjected to the test of time, including repeated challenges and refinements by competent critics, become the foundations of science.

Science defies simple definition for the same reason that most complex intellectual enterprises cannot be adequately described in a line or two; brief "definitions" may be meaningful to the expert but rarely tell the full story. Is science simply what scientists do, as some circular definitions would have it? Or is it a body of useful and practical knowledge about the universe? Or is it a method of inquiry? Or is it the search for order in nature? Or is it a search for "first principles"? Or is its objective to understand, explain, and make predictions about natural phenomena? Yes, science is all these things, yet more, for what is lacking in such simple statements is really the essence of science: the search for *verifiable* "truths." The scientist/philosopher Norman Campbell characterized science as "the study of those judgments concerning which universal agreement can be obtained,"[1] while the noted biologist George Gaylord Simpson chose to portray it as "an exploration of the material universe that is both orderly and self-testing,"[2] and the philosopher of science Ernest Nagel described its objective as follows: "The aim of the scientific enterprise is to provide systematic and responsibly supported explanations."[3]

These definitions may seem different, but in the broader sense they are not; taken in proper context the terms "universal agreement," "self-testing," and "responsibly supported explanations" mean much the same thing. First, scientists may hold personal views on how best to define science, but their opinions on the workings of nature must be able to withstand the scrutiny of their peers. Not surprising, as the same may be said of any intellectual discipline, whether in the humanities or the sciences; works of art, poetry, literature, music, philosophies, religions, and so on cannot long survive without "universal" (i.e., widespread) agreement as to their merits. But the distinctive feature of science is that in the final analysis it is the *only* discipline in which the truth or falsity of one's statements about its subject matter, namely nature, can in principle be tested *objectively* by putting the questions to nature itself, that is, by performing properly designed experiments. This is the glue that holds the enterprise together. The scientific community has a reasonably clear standard of scientific "truth," one based on careful observation, experiment, logic, and, above all, replicability. For its continued progress, science depends upon such "truths"; and one really cannot appreciate what science is about without first understanding this meaning of truth, which is not necessarily the same as the common notion of truth in everyday life. Not even do the so-called facts of science become facts simply because of random observations (which may be inaccurate), but only after repeated analysis and confirmation.

Our normal commonsense impulse is to think of "truth" as something that *is* known. This is too restrictive a definition to apply in science, however, where the epistemological definition of "truth" as what *can* be known (sometimes stated as "what *can* be verified") is more appropriate.[4] That is, since most useful statements in science are essentially predictive (e.g., given a particular set of conditions certain consequences *will* obtain), they are incapable of being *completely* proved. Instead, the laws upon which such predictions are based must be taken as provisionally *true* if many instances of their truth have been found and none of their falsehood. For example, the reader would surely attest to the a priori "truth" of the statement: the Sun will rise in the east tomorrow, and probably also to the prediction that Halley's comet will return in 2084. The reason for such beliefs is that experience convinces us of the regularity of planetary motions. But is Newton's law of universal gravitation, which provides the scientific basis for such predictions, equally true? The answer here is "yes," with a very high degree of probability, at least as far as its explanation of simple planetary motions is concerned. More will be said later on the concept of scientific truth.

▲   NATURE IS ORDERLY

The practice of science is predicated upon the belief that at least in the *macroscopic* sense nature is orderly and predictable, that is, under the same conditions the same phenomena will be observed, that the Sun, the moon, and the tides will behave as expected, as will all natural phenomena of which we have cognizance. One might call this an article of faith, but centuries of experience lend support to such a tenet. The main objective of science is to find a common basis for understanding our encounters with nature. To accomplish this the scientist generally devises models or conceptual schemes that provide his or her peers, that is, the scientific community generally, with reasonably convincing explanations. Note that the emphasis here is on acceptance of new ideas by the scientific community rather than by society at large. The reason is that unlike early science, when explanations generally could be couched in terms meaningful to the layperson, modern scientific understanding does not necessarily accord with the commonsense beliefs of the average individual. Science begins in the "real" world with observation, and in the end often returns to that world through technology in the form of tangible products. The beginning and end are easy for the nonscientist to grasp because they represent reality, but in between, where the cognitive part of the process occurs, is often too remote from one's notions of common sense to be readily accepted by the general public. For example, consider an object resting on a table and ask how one accounts for this simple phenomenon. The average person would hardly think this warrants a scientific explanation because it is perfectly obvious that the object is being "supported" by the table. But this is not a scientific explanation; it is simply another commonsense way of saying that the object is on the table. Newton would tell us that the object pushes down on the table and therefore the table must push back on the object. Clearly there are forces involved, but of what sort?

Giving a full scientific explanation would require first invoking the atomic theory of matter, which has both the object and table consisting largely of empty space with their masses concentrated in the nuclei of atoms that are situated at great distances from one another compared to their nuclear size.[5] So why does the object not fall into the empty spaces of the table? It would, if the object were a sharp needle being pushed into the table. But otherwise our model of the atom tells us that electrons move about in the empty spaces between the nuclei and that these electrons, being negatively charged, repel one another so that the object essentially "floats" on a sea of negative charge at the surface of the table.[6] The needle penetrates because, being sharp, it cannot carry enough charge at its point to repel the table. Then, as if this explanation weren't bizarre enough to our commonsense individual, we call upon the very same electric charges to hold the molecules of the table together, as well as those of the object, in the form of rigid structures.[7]

Were it possible for scientists to communicate these thoughts and actions in common language, using models to provoke images that are as meaningful to the nonscientist as to their peers, they would happily do so. But the nature of modern science, particularly those branches that have matured to the point where their frontiers now involve much deeper subtleties than they did a century ago, makes this impossible. After all, there is no good reason to believe that our everyday world, the world of ordinary sensations, modes of thought, and descriptions, is the same as that in which best explanations can be found for the most elusive aspects of nature. Hence, it is not surprising that many of the ideas of modern science seem so strange to the average individual, who is unaccustomed to thinking in a scientific fashion. Science, as we know it, could not progress if it were bound by the requirement that all its statements make sense to the untrained individual. Hence the scientific community requires only that new ideas be acceptable to one's fellow scientists rather than to all members of society. Consider a simple analogy: you wish to attend a lecture in a foreign language because the subject interests you. If you do not understand the language you might do one of two things: you could first learn the language, which would be impractical, or you could have it translated. In science, however, "translation" into a more familiar language is not the answer. As an interested observer you would first have to learn a reasonable amount of science in order to fully appreciate a seemingly simple explanation such as that given above for the object resting on the table. It cannot be translated for you in popular terms without losing much of its real meaning. The scientific descriptions of the object and the table are so vastly different from the way they appear to the commonsense individual as to defy simple, everyday explanations. Indeed, some philosophers of science, like Ernest Nagel, would assert a basic flaw in the "two tables" example, namely, that even the languages of experimental and theoretical science cannot be equated; that the word "table" is an experimental concept not found in solid state or atomic theory; and similarly that the word "atom" is a theoretical idea not defined in the language used to describe observations or experiments. Hence they cannot be intertranslated.[8] Epistemological considerations aside, however, the example does point up the difficult problem of communicating science in commonsense terms.

▲ THE SEARCH FOR TRUTH

Science is often described as "the search for truth" about nature, but this phrase implies absolute truths, a concept that demands caution. Yes, science does seek the truth, but does it always get "the whole truth and nothing but the truth," as goes the well-known courtroom oath sworn to by witnesses? One may question the meaning of truth, scientific or otherwise, on purely philosophical grounds, but here we take the pragmatic view of equating it with success, that is, laws or theories are regarded as "true" only to the extent that they are successful in accounting for the particular segment of nature claimed for them. As Bronowski put it: "Science is a great many things, . . . but in the end they all return to this: science is the acceptance of what works and the rejection of what does not. That needs more courage than we might think. . . . It needs more courage than we have ever found when we have faced our worldly problems."[9]

Because of science's tentative character, it should be evident that we cannot claim *absolute* truths in science, so how much of the ultimate truth we actually know depends upon how well we phrase our questions of nature, that is, how carefully we design our experiments. Bear in mind that a single negative finding (null experiment), if confirmed, is sufficient to disqualify a scientific hypothesis or theory. On the other hand, repeated verifications of a given theory only add to the confidence the scientific community has in its validity and diminishes the probability that a null finding may turn up—but it may not tell the whole truth. That is, it may be supplanted by another theory that allows for more accurate prediction or covers a greater range of application. This is not to say that the original theory was wrong, but that it was useful only within certain prescribed limits. Whenever one deals with a scientific law or theory its limitations ("boundary conditions") must be kept in mind.[10] For example, Newton's well-known laws of motion must be modified under certain conditions according to Einstein's principles of relativity, and the classical gas laws strictly apply only to so-called ideal gases and must be adjusted both for very high and very low temperatures and pressures.

Errors are made in science that go beyond mere incompleteness, but in the normal process of self-correction even these are usually revealed. A good example here is the structure of the atomic nucleus as it was envisioned in the 1920s and early 1930s, before the discovery of the neutron in 1932. At that time, and even in the 1930s and 1940s, before the neutron could find its way into high school textbooks, most chemistry and physics texts had protons and electrons as constituting the nucleus of the atom, which accounted pretty well for much of what was then known about nuclear properties. There was plenty of speculation, of course, on the possibility of a new particle, having no charge, and a mass slightly greater than that of the proton. But not until the existence of the neutron was confirmed experimentally could the correction be made with some confidence, leading to a more satisfying picture of the atomic nucleus and opening still another major field of study in nuclear physics.

Hence, science in a sense is tentative and provisional; but this should not be taken to mean that the entire edifice can crumble like a house of cards, for science is also cumulative. Its structure has been built up over the ages, but always of removable bricks, so that the weaker ones, when discovered, may be replaced, leaving a foundation that grows stronger with time. Sometimes observed facts (anomalies) turn up that cannot be explained by current theories. Such challenges inevitably stimulate a fresh look at the existing conceptual framework and lead to new conceptual ideas that are more inclusive than the old.[11] It is this cumulative nature of science, according to Conant,[12] that distinguishes it most from other forms of intellectual activity. He points out that while progress is expected in the sciences, it is not so much a requirement of the humanities. The difference actually goes much deeper. One may question, for instance, whether it is even meaningful to speak of progress in the humanities. A more useful, though related distinction, may be the question of self-consistency with respect to methodology. Science, we have said, has evolved means of testing the "truth" of its statements about nature. These methods, while not infallible, have generally proved successful in guiding science along the path to useful knowledge. Having a path of continual self-correction should be appreciated as the single most powerful feature of the scientific enterprise. Nash terms this the "principle of corrigible fallibility," pointing out how it enables scientists to explore even the most far-fetched ideas:

> However we conceive science, we cannot begin our work without some facts and some ideas provisionally accepted as unchallengeable. But the ideas may be quite wrong; the facts ill-observed or laden with 'credulity.' Obsessed by such fears we don't do science! From such obsession scientists are freed by what I call the principle of corrigible fallibility. It is a principle of action. Beginning with the best available facts and ideas, we proceed vigorously in the faith that any errors in them will be revealed by the *interaction* of facts and ideas—by the *interaction* of rational and empirical elements, in neither of which we have, or can have, absolute confidence. We amend our hypotheses in the light of our experiments; but we also reject (as errors), 'correct,' and explain away some of our data when they conflict with 'indubitable principles.' In such unquestioning acceptance of principles to which all experience is made to conform . . . science may seem at one with divination, or magic. Yet science progresses, as magic does not, and not simply because science had the good fortune to hit on the 'right' principles at the outset. It did not! But it *learned* better principles.[13]

On the other hand, despite frequent claims to *self-evident* truths, the humanities have no logical rules or methods for testing their progress. Indeed, the notions of truth or falsity cannot even be applied to most humanistic knowledge; it is meaningless, for example, to question whether a work of art or of poetry can be justified solely on such grounds.

▲ THE CONCEPT OF FALSIFIABILITY

A further word is in order relating to null experiments. We have seen that the essence of natural science is self-correction, that every statement must be verifiable; but is corroboration alone sufficient? Some empirical statements about nature are easily verified; for example, under specified conditions, reacting element A with element B *always* produces compound C. Although seemingly trivial now, such a statement would not have been so regarded in the seventeenth and eighteenth centuries, when the atomic theory of chemistry was in its infancy. But while the statement in question may be deduced from a grand conceptual scheme (the atomic theory), does its verification also establish the "truth" of the parent theory? Not so; it lends support to the overarching theory, but one cannot extrapolate from the accuracy of singular (or even many) deductive statements to the correctness of a theory that was arrived at by induction, that is, by generalizing from a limited range of experience, or even by intuition, imagination, or educated guesswork. There is always the possibility that an exception will be found that renders the theory false. As philosopher Karl Popper points out: "no matter how many instances of white swans we may have observed, this does not justify the conclusion that *all* swans are white."[14] However, except for the possibility of a gross sampling error, it does suggest a high probability that all swans are white.

Virtually all scientists take the position that repeated independent testing and verification of a given statement about nature are sufficient to establish its validity. And most would also agree with a central philosophical concept of the theory of knowledge, sometimes called the "problem of demarcation," that all *meaningful* scientific statements must be both verifiable *and* falsifiable, that is, the statements must be of such form that both tests are logically possible. The criterion of falsifiability is invariably applied in the practice of science, sometimes in the process of verification but usually—and often subconsciously—in the evolution of scientific statements, whether these be called postulates, theories, laws, or any other name. Thus, logical positivists hold that it is not sufficient for a statement to be subject to positive verification but that its form be such as to be falsifiable as well, to be contradictable in principle by experience. Falsifiability, although rarely used in everyday affairs, provides a means of testing the *logical* validity of all statements, whether in science or otherwise, provided that the statements can be phrased precisely enough so that one may contemplate how they might be falsified. Thoughts relating to qualities rather than things or happenings, for example to feelings or sentiments, or to human characteristics or behavior, are often so vague and equivocal that they cannot be put to this test. Stating that an individual is honest is quite different in this sense from stating the individual's height or weight. Honesty lacks precise definition and hence cannot easily be verified or falsified, while one's physical characteristics obviously can. Thus, while falsifiability might be a desirable logical test in many instances, it is unfortunately not readily applicable to everyday life.

However, falsifiability is an excellent criterion for identifying pseudoscientific claims, which typically are so stated as to defy definitive experimental test. The fact that claims such as extrasensory perception (ESP) or other paranormal phenomena are generally not reproducible is one test, but as pointed out by Lawrence Cranberg, there is a "deeper theoretical objection" to such claims, which is that they are not falsifiable.[15] Cranberg gives the example: "You can drain Loch Ness through a fine filter without proving that the Monster was never there." Consider some other simple examples. The statement "it will rain in New York before noon tomorrow" is falsifiable, whereas the statement "it will rain or not rain in New York before noon tomorrow" is logically incorrect because while it is verifiable, it is not falsifiable, that is, the statement cannot be refuted. Not all either/or statements are necessarily invalid, however. Thus, while the statement "it will rain *or* snow in New York before noon tomorrow" is not falsifiable because it lacks precision, the corresponding statement, "it will either rain or snow in New York before noon tomorrow, but not both," is perfectly valid. The strength of pseudoscience, such as it is, rests on the premise that while its statements cannot be proved correct, nor can they be proved incorrect. Asking one who makes such a statement, "Can you prove it?" is likely to be countered with "Can you disprove it?" Falsifiability provides just such a test. For example, a claim such as "the Earth is flat," while not easily proved false at the time of Columbus, was falsifiable in principle—one could devise an experiment, such as the voyage undertaken by Columbus, that would disprove the claim. Thus the hypothesis of a flat Earth was scientifically *meaningful*, even though incorrect.

Astrology, among several other equally mindless customs, is noted for its pseudoscientific assertions. A typical horoscope for Aquarius (January 20–February 18) reads: "Aquarius is assertive, independent, progressive, analytical, original and inventive. They have strong dislikes and opinions." This statement is not falsifiable, but could be if there existed absolute measures of assertiveness, or independence, or progressiveness, so that a given Aquarius could be shown to lack one or more of these qualities. Further on in the same horoscope appears the statement: "Aquarius are prone to ankle sprains and breaks, also to varicose veins and hardening of the arteries." Such a statement, lacking qualifying adverbs like "more" or "less," is also nonfalsifiable, and is therefore meaningless. This is characteristic of virtually all pseudoscientific claims. Either they are purely subjective and nonreproducible, such as "a UFO landed in my yard last night and then took off again"; or so loosely framed as to be nonfalsifiable, like the astrology examples cited above. The few that are falsifiable, such as the flat Earth theory or the tenets of creation science, are so readily proven false today that they might better be termed "unscientific" rather than "pseudoscientific." Cranberg makes the further useful observation that when dealing with such terms as "unnatural" or "paranormal" or "extrasensory," we must face up to the lack of precise definition of the terms themselves. That is, at what point does an observation become paranormal or extrasensory? Scientists generally believe they can tell the difference between normal and paranormal, and probably they can in most instances, because these are generally so obvious, but it would be helpful to have clear-cut "scientific" definitions of such terms for borderline situations.

▲ WHAT SCIENCE IS NOT

Having set down in aggregate what science is, it may be instructive to say what science is not. We have already pointed out that science is not merely the process of collecting facts. Science begins with observation, to be sure, and careful observation means acquiring factual data, but this is a means and not an end to the quest. As the famous eighteenth-century English clergyman and chemist Joseph Priestley said: "The first object of physical science is to ascertain facts."[16] Granted that judgment is required in deciding which facts to collect since it is fruitless to study nature at random, but simply gathering data and even looking for similarities among related phenomena so as to classify them according to certain accepted rules, sometimes called "natural history," are only a small part of the enterprise. A useful maxim to bear in mind here is that *nature study* is not the same as the scientific *study of nature.*

Nor does science consist of a "tried and true" routine for acquiring knowledge of the universe. The seventeenth-century philosopher Francis Bacon, most of whose early visions of the social utility of science actually came to pass, believed (incorrectly, as it turns out) that science should be practiced by making all possible observations, gathering data, and then, by systematic induction, generalizing to scientific laws.[17] Bacon was not a scientist and his system failed to work even then, when science was more descriptive than it is now; nevertheless, he led the revolt against medieval church-centered learning, and his attempts to systematize the practice of science profoundly influenced his successors, leading many to seek ways of describing the process in operational terms. It became fashionable early in the present century to speak of *the* "scientific method," as though there was a set procedure for doing science which, if followed, was guaranteed to produce results. Its purpose was not to establish a recipe for scientists to post on their walls but as a guide for students on how scientists practice their discipline. Influenced largely by the writings of the English biometrician Karl Pearson at the turn of the century, most elementary science texts in the first half of the century included some discussion of scientific method.[18] Generally this consisted of variations of the following steps: (1) a statement of a problem is made; (2) relevant observations are made; (3) a hypothesis is formed; (4) predictions are based on the hypothesis; (5) an experiment is performed to test the predictions; (6) the hypothesis is accepted, modified, or rejected.

We have here an apparently simple methodology, whose elements can be discerned in one form or another in most scientific research, and one that seemingly lends itself well to science education. But the problem with it is that, except for the most straightforward situations offering clear cause-and-effect relationships, the notion that here is a routine for automatically solving any scientific problem is patently false. The most glaring omissions are knowing what the pertinent problems are, what observations should be made, and how to be sure that the experiment performed unequivocally tests a given hypothesis. In other words, the question is not so much *what* steps are normally performed in a scientific investigation, but *how* one goes about performing them. As any practicing scientist knows, this cannot be formalized; science is a "no-holds-barred" struggle with nature, imagination and the prepared mind being the main ingredients for real progress. The issue was well characterized by the physicist Max Born when he wrote: "I believe there is no philosophical highroad in science, with epistemological signposts. No, we are in a jungle and find our way by trial and error, building our road *behind* us as we proceed. We do not *find* signposts at crossroads, but our own scouts *erect* them, to help the rest."[19]

▲ SCIENCE VERSUS COMMON SENSE

Nor, as we have seen, does science necessarily accord with common sense. In the latter part of the nineteenth century, T. H. Huxley, seeking to persuade the general public that science was not a form of black art but rather a kind of common sense, gave as his argument the fact that both scientists and the public use inductive and deductive reasoning in their everyday activities.[20] "Thus," he said, "one should not conclude that science has a special way of looking at things that is not also the way of ordinary people, except that scientists must be somewhat more precise and take greater care in verifying their conclusions through appropriate experiments." He was only partly right then and would be less so today; the two use similar forms of reasoning and often deal with everyday things, but where they differ so markedly is in how they apply such reasoning: everyday things make "sense," while many of the models of modern science do not. The fact is that modern science (and mathematics) and the kind of reasoning that is characteristic of science are remote from one's everyday experience. Until roughly the end of the nineteenth century, when the universe could still be described as a well-behaved, deterministic mechanism of moving "billiard balls" (atoms/molecules), it was possible to talk of it in common-sense terms; at least one was not required to stretch one's imagination beyond reasonable limits. But then the picture changed. Our most basic concepts, upon which in the final analysis we build all of science, became so abstract as to defy simple, commonsense understanding. To make matters worse, from the point of view of the general public, determinism in science gave way to probability, and the universe might now be viewed, in a sense, as a "cosmic game of chance."

Huxley's attempt to fit science into the mainstream of everyday life illustrates the tenacity with which we all tend to cling to common-sense descriptions. What David Hume, in 1740, called "natural instinct" and Thomas Reid later termed "common sense" is, in fact, the guiding principle for most of what we think and do. Reid maintained that the "consent of ages and nations, of the learned and unlearned, ought to have great authority," especially so when the principles involved "concern human life" and are the basis of "our ordinary conduct of life."[21]

We know that in everyday life common sense, or "natural instinct," easily triumphs over philosophical doubts, and this, it appears, is what often happens when we try teaching modern science to the average individual. We have no commonsense counterparts for many of the things we talk about in science, yet students demand concrete explanations that are meaningful to them. The molecular theory of matter, the kinetic theory of heat, the gene theory of heredity—all deal with concepts that are not directly accessible to our senses. The lack of meaningful models is a great disadvantage, of course, when dealing with students (and particularly adults) who are bound by commonsense experiences and tend to reason by analogy with such experience. There is another aspect to the question of science versus common sense that has to do with the methods one uses to assess the validity of our statements. As pointed out by Ernest Nagel, it is the preciseness of scientific statements that strikes the commonsense individual as being so different from his everyday experience.[22] One of the features of common sense information—information acquired in the course of one's ordinary experience—is that while the information may be accurate enough within certain limits, it is rarely accompanied by any explanation of why the alleged facts are believed to be true. And when "common sense" does try to provide explanations for its facts, the explanations generally fail the critical test of relevancy, but stem instead from folklore, hearsay, or pure speculation. Thus, it is certainly true that a falling stone accelerates toward the earth, but not because of the Aristotelian explanation that "the traveler hastens as he approaches his destination." To underscore this point, Nagel suggests that commonsense knowledge frequently deserves the advice given by Lord Mansfield to the newly appointed governor of a British colony who was unversed in law. Mansfield said: "There is no difficulty in deciding a case. Hear both sides patiently, then consider what you think justice requires, and decide accordingly; but never give your reasons, for your judgment will probably be right, but your reasons will certainly be wrong."

Even practicing scientists frequently find themselves bound by their commonsense experiences. For example, the modern cognitive/ computer scientist Douglas Hofstadter is reluctant to give up on common sense as the basis for doing and understanding science, but he clearly recognizes the difficulty of this position. In a recent address he asserted that "science is just a highly developed form of common sense. To justify that would be difficult, but this is my view and I think that most of those in this audience share it."[23] However, in picturing common sense, he added that it was "murky, complex, and troubling, with great potential for error making due to the complexity of the underlying concepts." Common sense, as Hofstadter sees it, is a "bull's-eye of rationality surrounded by a blurry extension of the concept." One wonders whether recurrent attempts to identify science with common sense may not be conditioned more by what the proponents believe common sense ought to be than what it actually turns out to be in practice.

Individuals accustomed to commonsense explanations find the transition to the systematic explanations of science somewhat disquieting. Most are surprised to learn, for example, that scientific theories, which somehow seemed endowed with everlasting qualities, rarely have the permanence of commonsense beliefs. And the reason, we know, is that the greater precision demanded of scientific statements renders them so much more subject to change in the light of repeated challenge by experimental evidence. Commonsense statements, on the other hand, being generally of the kind not confronted by painstaking evidence, have such wide latitude as to endure often for centuries. The commonsense individual may be further surprised to learn that while the precision of scientific statements exposes them to greater risk of being found in error than loosely worded commonsense statements, the former nevertheless have the great advantage of being more readily incorporated into comprehensive, clearly defined systems of explanation, which systems play the leading role in our scientific enterprise.

Huxley apparently was not quite as wedded to the notion that science and common sense are equivalent as might appear from his efforts to convince the public of it. He says, for example, in "The Method of Zadig," that "science is nothing but trained and organized common sense, differing from the latter only as a veteran may differ from a raw recruit: and its methods differ from those of common sense only as far as the guardsman's cut and thrust differ from the manner in which a savage wields his club."[24] The telling phrase is "trained and organized common sense," which, we must admit, is an uncommon sort indeed. What Huxley failed to point out in his metaphor, however, was the self-testing feature of science, absent in most ordinary reasoning except the very trivial, that makes it possible to invent the strangest models and conceptual schemes, often quite opposed to "common sense," and yet after being tested against nature they may prove the best means available of representing the real world. The commonsense individual has no such equivalent. If one generalizes by induction from a false premise, as so often happens in daily life, the generalization is equally false and often cannot be corrected; or if conclusions are improperly deduced from false or incomplete evidence, there is no way of knowing that the conclusions may be false. It is difficult to convince the commonsense individual that electrons, neutrons, atoms, genes, chromosomes, black holes, and the like are not intended by science to accurately represent reality, but are only mental images (constructs) created by science as a convenient means of portraying portions of the real world. Several years ago, Daniel Boorstin, the Librarian of Congress Emeritus, acknowledged the difficulty of closing the huge gap between the citizen's desirable and actual understanding of science: "While understanding science is more urgent for us than ever before, achieving that understanding seems less and less attainable. We are baffled and dazzled today by new concepts and 'entities'—from double helixes to black holes—that defy common sense, yet we are still expected to have an opinion on what to do about them."[25]

Boorstin hit at the heart of the problem, for we have seen that so much of modern science defies common sense that one should not be surprised by our failure to achieve universal scientific literacy, particularly in the sense of civic responsibility. If science is to be equated at all with common sense, we have seen that it would have to be with such a highly organized and structured form of common sense as to appear no less strange to the average commonsense individual than does science itself.

▲ NATURE APPEARS ORDERLY

Were it not that nature seems to work in essentially simple fashion there could be no science. Suppose, for example, instead of having roughly one hundred different chemical elements in nature there were 100 billion. Then, as Poincaré pointed out,[26] the chance of finding two objects in nature bearing some resemblance to each other would be so remote as to make it impossible to find any sort of intelligible pattern. Whatever we might know of parts of the universe would avail us little in our efforts to understand the whole. All things would be different and science could be no more than an interminable collection of isolated facts. We are accustomed to finding diversity in nature; fingerprints are unique to an individual, as is the individual's DNA, and differ in principle from all others. Individual features also differ, usually making it possible to distinguish one person from another by appearance or speech patterns. But not surprisingly, considering the evolution of the species from a single archancestor, all human organisms have architectural similarities that transcend their intraspecies differences and distinguish them from other species of the animal kingdom. These similarities make possible, for example, not only the taxonomic classification of organisms but also the practice of medicine, which clearly could not exist if every organism differed from every other one. Indeed, under such conditions civilization itself could not exist.

By the same reasoning, however, nor could there be any meaningful art. Nature would defy description by the humanist as well as the scientist. Neither one could evoke memories of the past nor provide useful ideas for the future. Imagine that man's range of hearing or the spectral limits of his vision were multiplied a thousandfold, and think what effect this would have on music and the visual arts. That order and simplicity are prerequisite to all forms of knowledge, not only science, must be evident. Given two alternate scientific explanations for the same phenomenon, the better choice should invariably be the one offering the greater economy of thought and expression—sort of the scientific equivalent of a seldom used but highly important principle known as "Occam's Razor."[27] Occam's Razor admonishes us to favor the simplest explanation that accounts for the observed facts, which has a commonsense ring to it. Deciding which of several alternative explanations is the simplest is not always easy, but on the whole this precept has proved of inestimable value to science.[28] Attractive as this guiding principle may seem, however, it should not be taken as an inviolable rule. While convoluted explanations are to be avoided in science, as in all serious discourse, it does not necessarily follow that parsimony alone leads to more precise explanations.

Not only do we assume that nature is compliant but that it is also orderly—that, given like conditions, scientific laws developed here on earth (or by some other civilization) are valid throughout the universe. The development of natural laws depends upon such regularity. Clearly we could not have one set of laws that apply on earth and another that apply on the moon, for example; what is true of parts of the universe must be true of all. Newton put this well when he conceived of universal gravitation, the first great inductive generalization in science:

> [A]nd in the same year I began to think of gravity extending to the orb of the Moon . . . and having thereby compared the force

requisite to keep the Moon in her orb with the force of gravity at the surface of the earth, and found them to answer pretty nearly. . . . All this was in the two plague years of 1665 and 1666, for in those days I was in the prime of my age for invention and minded mathematics and philosophy more than at any time since.[29]

That Newton's generalization was correct has since been amply demonstrated through astronomical observations, and in recent years by the precision of space flight calculations. An interesting philosophical note might be added here on the universality of scientific laws. Poincaré asks whether Newton could have arrived at his law of gravitation had there been no astronomical observations at the time, had the Earth, for example, been forever blanketed by a cloud cover much as Venus is.[30] Never mind the problem of how we could get our energy from the Sun or how life would have evolved under such conditions; purely as a hypothetical question, how would science have developed without our awareness of the planets and stars? Could Copernicus, Kepler, Galileo, and Newton have made the progress they did? Could Newton, in particular, have devised his law of *universal* gravitation from the famous "apple tree" observation alone? The answer is that one does not need to know that the Earth is part of a much larger universe to develop physical laws that apply on earth (and certainly not for Earth-bound sciences like the life sciences), but it was astronomy in the first instance that taught us about regularity in nature through the motions of planets and the apparent motions of the stars. So, argues Poincaré, it was our ability to observe the heavens that convinced us that the universe was governed by natural laws and the challenge to discover these laws was the start of science. Hence it is very unlikely that without the astronomical work of his predecessors, Newton could have arrived at his law of gravitation, universal or otherwise. Methods for measuring the small gravitational attraction between laboratory-size masses were not available until a century later, when Newton's law was first verified experimentally by Henry Cavendish.[31]

A final point should be made with regard to extrapolating from limited experience to universal laws. One must be certain that the part of the universe under study is truly representative of the whole. For example, could Robert Boyle have deduced his well-known law of gases, which depends on the statistical regularity exhibited by large numbers of molecules of a gas, had he done his experiments in outer space (or in a largely evacuated vessel) where he would have had only a relative handful of molecules in his experimental chamber instead of the many billions upon billions that he did? Obviously not, for while the volume of the chamber is easily determined, the random fluctuations of pressure brought on by so small a sample of molecules would make it impossible to establish the pressure-volume relationship discovered by Boyle. Indeed, the concept of pressure in terms of a force exerted by a gas on the walls of its container would be meaningless and have to be defined in some other way. Newton did not have this problem; both the earth and the Moon became his laboratory.

▲ NATURE APPEARS DISORDERLY

Having asserted that science is possible because nature appears simple and orderly, it may be disconcerting to some that it appears orderly only because of the play of large numbers; that the laws of nature are valid only because of the statistical regularity of large numbers of events. On the microcosmic scale, nature is, in fact, random, and we are able to make sense of it because the "observables" we deal with, which according to some philosophers[32] are the only proper subjects of science, are generally very large. A period on this page, for example, contains on the order of a billion billion molecules; hence it appears stable and predictable. If it contained only a few tens of molecules, however, its behavior, like that of Boyle's fabled experiment above, would be unpredictable. So goes another count against science in the public mind; on the one hand science is presented to people as a way of understanding nature, while on the other they are told that nature appears on the submicroscopic level to be like a huge game of chance, with atoms and molecules as chips, and that the mere act of trying to predict their individual behavior with high precision is rendered uncertain by the observation itself.[33]

▲ THE LANGUAGE OF SCIENCE

Mathematics is the language of science, particularly of those branches of science that have evolved beyond the early descriptive stages. It is the only language we have by which statements about nature can be combined according to *logical* rules—a language that not only allows us to describe in precise terms the world about us, but also provides us with a means of dealing with such descriptions so as to lead to new knowledge. It is an essential part of the structure and practice of science, not merely an accessory. The physicist/philosopher Percy Bridgman once pointed out "that the fundamental human invention is language, and that we owe the progress of the race to it more than to anything else."[34] Similarly, mathematics may also be regarded as a language, and we probably owe the progress of science to it more than to anything else. When Galileo developed his science of uniformly accelerated motion (constant acceleration), he started with two definitions (of velocity and acceleration), which are easily stated in ordinary language as (1) velocity is the rate of change of distance (or position), and (2) acceleration is the rate of change of velocity. But since we have no formal rules for combining these two statements to form a third that provides new information, there is not much that can be done with them except to record velocity and acceleration according to these definitions. On the other hand, writing them in mathematical terms gives us the ability to combine the two statements (now in the form of simple equations: $v = \Delta s/\Delta t$; $a = \Delta v/\Delta t$) according to established mathematical rules to yield new relations among the four variables, providing thereby a means of testing the reasonableness of the original definitions. Here, Δs represents the distance covered in time Δt, and Δv is the change in velocity during that time. Suppose, instead, that the statements were simplified by writing them in a shorthand notation (say Pitman or Gregg), which some students mistakenly believe is the only purpose of mathematics. Again, it should be evident that little would be gained by this unless the shorthand method also provided rules for combining two or more such statements.

Contrary to the popular notion held by most students, the role of mathematics in science is not simply as a tool for solving numerical problems, although it is obviously useful in this way. It is mainly in the role of a *formal* language that science relies so heavily on mathematics. By and large, mathematicians do not care much for the depiction of their discipline as the language of science, which tends to confer upon it an adjunct status. They look upon its role as much more than this, and they are correct, of course. Mathematics does not consist solely of special techniques for computation, nor is it merely a useful language for scientists. Instead, it is a special kind of intellectual dialogue in which the ingredients are symbolic structures rather than common words, and in place of ordinary syntax the rule is to apply to these structures the most rigorous forms of reasoning that humans have managed to devise. It must be recognized that mathematical systems are created by man, not found in nature. A mathematician may choose to look at nature and try to formulate systems that relate to the real world. But in the main, mathematics is more concerned with the internal consistency and generality of its structures than with conforming to everyday experience. Thus, a particular mathematical system may be wholly abstract, its basic symbols assigned no explicit meaning and having no apparent connection with anything in nature. It is this proclivity for abstraction, toward exploring the full depths of human thought, that places mathematics at the very highest level of one's power to reason.

It is chiefly in this respect that science differs from mathematics. Like mathematics, science is also a brainchild of the human intellect. The so-called laws of nature are created by individuals, often after considerable struggle, not simply handed to us by nature. The models and conceptual schemes that are devised to depict the real world are the products of human reasoning and imagination. But while mathematicians are free to invent any logical structure they choose, even the most fanciful one, science in the end must conform with reality. It is this ultimate appeal to nature that sets science apart from all other forms of human activity. Thus, by resorting to experiment one has a means of testing the validity of any assumptions made about nature; in fact, even of probing the structure of science itself.

Science and mathematics share one common characteristic that we might wish otherwise; of all forms of organized knowledge they are the least understood by the educated public generally. One of the eccentric myths of modern times is that we live in a "scientific age." This may be true if judged by the massive technology surrounding us, and perhaps even by the major financial support given to science, but certainly not if measured by the scientific and mathematical sophistication of our society. The fact is that we live in a humanistic society. Our thoughts and actions are shaped much more by a humanistic culture than a scientific one. The general public responds more to emotion than to reason, to fantasy more than to fact, and to words more than to ideas. In short, our culture leans more heavily on its senses than its mind.

▲ CLASSIFYING THE SCIENTIFIC ENTERPRISE

Knowledge of the natural world falls into three broad categories: natural history, science, and technology. Science to the average citizen generally means the first and last of these, because it is through natural history and technology that most people are in touch with nature.

Among our earliest impressions are the natural phenomena of our environment, the kind that are seemingly simple and can be presented in purely descriptive fashion, descriptions that are meaningful because they correspond directly with one's everyday observations and senses. Observation, description, and systematic classification are the principal features of the division we have termed "natural history." It is the obvious starting point in a science curriculum, but it should not be the endpoint as well, as is unfortunately the impression of those who believe that science consists mainly in collecting facts. Science, we have already seen, seeks to account for the observed facts, while technology, which is more fully discussed below, seeks to adapt nature to society's needs. Collectively, these three divisions make up the overall scientific enterprise.

▲ THE ROLE OF TECHNOLOGY

Science is one member of an important symbiotic relationship in the total scientific enterprise; the other is technology, which, by extending human capability, allows us to harness or modify nature to our needs. Technology is much the older of the two, dating back about 13,000 years to the beginnings of agriculture, when cultivation of plants (food growing) began to supplant food gathering. There was no science as we know it at the time, only accidental discoveries and trial-and-error refinements brought on by necessity. As the Stone Age passed into the Age of Metals, more sophisticated tools appeared and serious efforts to understand rather than merely control nature eventually emerged. When to date the beginning of science depends upon how one chooses to define it, but two periods stand out as having had perhaps the greatest influence on its development. One is the period of early Greek science (in the Golden Age of Greece, ca. 300–400 B.C.) marked by such figures as Plato, Aristotle, and Hippocrates, and the other the late Renaissance period, known as the "scientific revolution" (ca. 1500–1750), when giants like Galileo, Kepler, and Newton laid the groundwork for modern science. It was also in this period that the close working relationship between science and technology became apparent. Science needed instruments to probe the workings of nature, demanding greater precision and variety in such instruments as its search for understanding grew more sophisticated. This led first to the training of talented instrument craftsmen who responded to the specific needs of individual scientists, but eventually to a highly skilled independent industry developing its own instruments and related products for the growing scientific enterprise. However, this could hardly be considered the chief example of a science-technology interaction. New discoveries in science—in chemistry, metallurgy, optics, etc.—new inventions of a practical nature, some, like those in electricity and thermodynamics, stemming from science, others, given the range of materials and skills then available, from pure ingenuity, all led to the growth of entire industries. It is in this connection that one normally thinks of science as giving rise to technology. The feedback to science in the form of new instruments and materials, while obviously of major importance to furthering its development, is more a behind-the-scenes process that rarely captures the attention of the public.

The range of technology, sometimes known as "industrial science" or "applied science," is enormous—from the simplest consumer products such as paper clips and pencils to such huge utilities as electric power and communications; from mining ores to roads, buildings, and bridges; from making things easier for people to do manually to automated mass production techniques; and even to making some things possible to do at all ("new technologies"). Every individual is touched directly by technology, most heavily in industrialized societies, and western civilization could not have advanced so rapidly since the start of the Industrial Revolution (ca. 1760) without this enterprise. Not everyone agrees that this has necessarily been for the overall good of society; some social critics and poets deplore the inroads of technology on the tranquillity of modern life, about which we shall have more to say later.

There is an unfortunate tendency among some scholars to regard technology as being somewhat lower in the intellectual pecking order than pure science—that is, a discipline involving less mental agility—and while acknowledging that technology results in great practical benefits it does not produce the sort of universal conceptual schemes (theories) that are characteristic of science. This is a form of intellectual snobbery that is altogether too common when comparing the pure with the applied in many fields, both in science and in the arts. The view has long been reflected in standard science textbooks, which understandably, as they are usually guided, reviewed, or written by academic scientists, emphasize science over technology. Only in recent times do we find technology education movements gaining acceptance in our high schools and colleges. Yet technology, we know, is much closer to the student than is science, more meaningful because it deals with real things. As we shall see later, achieving some measure of technological literacy in our schools may be tenable, where scientific literacy is not. In short, technology has the "hook" that science lacks for most students.

To the average individual, technology is far more visible than science, yet somehow the two are usually lumped together under the generic "science." One of the puzzling (and discouraging) features of the current science education movement has been its failure to distinguish clearly between these two related but quite separate enterprises, having different motivations, objectives, and methodologies; yet in the public mind they seem to be regarded as one and the same. Not too long ago science and technology were treated as distinct subjects. *Science* was taught in our schools and colleges, not technology, at least not under that heading but often under the alias of "industrial arts." Now, probably as a result of the concerted drive during the past few decades to introduce technology education into the schools, they are often considered one and the same. The problem with this is that most of society experiences only the end products of scientific inquiry, namely those produced by technology, and almost all so-called science-based societal issues are actually based in technology rather than in science. Hence painting both with the same broad brush is a disservice not only to the science and technology communities, but also to society, which must understand that technology is fundamentally a social activity and that the social and economic forces which prompt technologists to modify nature are very different from those that motivate scientists to seek ways of understanding it. Whatever negative image the public may hold about science can usually be traced to its more direct interface with technology—to mass production, smokestack industries, and the like—which makes it all the more important to clearly distinguish the two enterprises. All thinking individuals must appreciate that while society is strongly affected by technology, society, in turn, can influence the directions taken by technology. Science is in a different category, however, for one can rarely forecast the uses to which scientific discoveries may be put or what the societal impact of a particular field of study may turn out to be. It is often said that we live in a scientific age, but as far as the public is concerned it is really a technological culture.

While in the physical sciences it is generally fairly easy to distinguish between science and technology, the issue is not always so straightforward in the biomedical sciences, where the border between the two often becomes blurred. A useful distinction is that science seeks to discover new knowledge of the universe while technology shapes that knowledge to human needs. But where, for example, should one fit the development of a new vaccine that used known experimental techniques or a series of clinical trials to establish the efficacy of a new drug? Based on the fact that both examples involve the development of practical end products, one might be tempted to consider them technology, but actually they could also be classed as science. The reason is that as the systems one deals with, especially living systems, become more complex, it becomes less evident that all the basic science needed to develop a given end product is already known; some new problems almost invariably turn up during the development stage that require further investigation. Thus, while clinical trials proper may indeed be an example of technology, the information gained from them may lead to new knowledge or at least point up gaps in existing knowledge. Hence some care must be exercised when categorically classifying an activity as *either* science *or* technology.

An essential feature of technology is its concern with risk-benefit assessments. Civilization has never lived in a totally risk-free environment; nor can it hope to do so today or in the future. In modern societies technology has advanced too far and society grown too much dependent upon it to turn back the clock. There is a school of thought that says any risk, however small, calls for absolute prevention rather than mere control, but this is clearly impractical in most cases. One thing we have learned from technology is the need to evaluate the benefits as well as the risks involved in the use of technology, to the extent that these can be determined, and predicate our judgments on the relative weights of such assessments. As the sixteenth-century political philosopher Machiavelli pointed out, "A man's wisdom is most conspicuous when he is able to distinguish among dangers and make choice of the least."[35] But equating every risk with danger is too simplistic, for while taking risks may be voluntary the harm that may result is certainly not intentional. In a sense everything we do carries some element of risk, the most obvious being crossing the street, or smoking, or traveling in automobiles, or living in earthquake, flood, or other disaster-prone areas. These are avoidable risks, yet they are accepted because many people believe that certain benefits outweigh their risks. The same reasoning should be sensibly applied to the risks of technology, whether to our health, our safety, or the preservation of natural resources. Risk-benefit analysis, something we do at least subconsciously in most of our daily actions, should be a guiding principle whenever society reflects on the possible ill effects of technology, certainly in those cases where it is possible to assign quantitative or even semiquantitative estimates of risk.

The real problem arises in connection with involuntary risks, which may be thrust upon individuals by smokers, reckless drivers, polluting industries in the neighborhood, and the like. Here the risk to the individual must be weighed against the benefit to others or to society generally, as well as against the cost of eliminating or reducing the risk. Science seems to have a "mandate" from society that permits it to engage in activities that may involve some risk, but society assumes that the risk will always be minimized. It is in this regard that risk-benefit analysis is gradually evolving into an effective regulatory tool. Sound regulations obviously increase public confidence that the system works. One of the problems in communicating such evaluations to the general public, however, is that the results are generally expressed in probabilistic terms. That is, instead of specifying a particular risk as being "small" or "negligible," it is more likely to be given as "one chance in a thousand," or "one chance in a million" of resulting in substantial harm. Such numerical data are meaningless to most individuals, who prefer to think in terms of zero involuntary risk or at most *negligible* risk. Unfortunately, despite the availability of a number of good published guides to quantitative risk assessment, the public seems uninterested in objective analyses, apparently preferring to be guided by their own intuition—or worse, by scaremongers. As long as innumeracy remains so prevalent in our society, the regulatory agencies will have to find more meaningful ways of presenting the results of their risk-benefit studies. It is often said that if the public is unable to evaluate the risks, it will evaluate the regulators; hence at the very least the public must understand the regulators.

▲   THE ROLE OF ENGINEERING

Just as there is confusion in the public perception of science and technology, there is a lack of understanding of the role played by engineering in the overall technology enterprise. Technology, we have seen, covers a huge span of activities, from doing or making the simplest things to the most complex. Everything done by technology must, of course, obey natural laws, but much of it is so rudimentary that it involves more trial and error than reasoned analysis. However, the more complex forms of technology, which could prosper only through painstaking design, eventually gave rise to the formal disciplines of engineering, whose main objective is to reduce the purely empirical content of technology, to make it, in effect, more "scientific." It is interesting to note that the first engineering society was the Institution of Civil Engineers (formed in early nineteenth-century London), which was concerned with controlling power sources; traffic on roads and rivers, including bridges, docks, harbors, and the like; the provision of water and drainage to cities and towns, machinery, and so forth— indeed, with the full sweep of the technology of the time. Actually, the first engineering school had been established in France some eighty years earlier, so that by the time the first society was formed the profession was already well regarded and in great demand. However, within a short time, because of the needs of a rapidly growing technology, specialization set in, and the first branch to be recognized as a separate discipline was mechanical engineering. This was not surprising, considering the importance of the newly invented steam engine and moving machinery generally. This was followed by mining engineering, concerned mainly at the time with locating sources of ores, coal, and industrial minerals. Then, as different areas of technology took on added importance, engineering fragmented still more into separate specialties, so that today one finds about a dozen separate departments in a typical engineering school. As might be expected from the lag between science and engineering, the first scientific societies were formed much earlier, in the seventeenth and eighteenth centuries, the most famous of these being the Royal Society of London, founded in 1660 for the purpose, as expressed in its charter, of "improving natural knowledge."

In a sense, engineering is the formal interface between science and technology. As science flourished, particularly during the last two centuries, it became too complex for the average craftsman, skilled though he might be, to translate its discoveries into technology; an obvious need emerged for a new class of technologists to perform this function, giving rise to the field of engineering. Engineers, who are trained in the sciences appropriate to their specialties, as well as in practical technologies, apply established scientific principles to improving the technologies and putting them on a more rigorous footing. Since most technologies involve interactions with society, engineers are also charged with protecting the safety of the consumer or user of a given technology. In brief, the engineer's job is to show us how to apply the "laws of nature" to improve our existence in the most efficient, most cost effective, and safest manner.

CHAPTER 4

The Scientific Literacy Movement

What science means and stands for is simply the best ways yet found out by which human intelligence can do the work it should do.
—John Dewey (*Science Education,* 1945)

▲   The current state of science education, as well as most of its problems, is clearly influenced by a single overriding premise: the *primary* function of formal science education, whether precollege or college, is to ensure a steady supply of scientists and science-related professionals, including, of course, science educators. Everything else that is done in science education, regardless of its educational worth or numbers of students involved, turns out to be secondary to this goal. Not that this is its avowed purpose, of course. Science educators persistently try to persuade themselves, and the community at large, to believe that there is a loftier purpose to science education, namely, to educate the general public—to achieve widespread scientific literacy— and indeed many educators have visions of such an ideal. But the reality behind such grand objectives is that the practical goal of producing future scientists must (and does) come first. Urging students into science as a profession is clearly part of the responsibility of science educators, but one must bear in mind that here we are dealing with only about 5 to 10 percent of the high school student population. To face the issue squarely, it is obvious that science departments, whether in our schools or colleges, are no different from most other disciplines in seeking to increase the enrollment of nonmajors in their courses; their underlying motive is more to attract critical masses of faculty and adequate equipment budgets for research than to satisfy some *compelling* educational need of the general (nonscience) student.

The competition is keen in most faculties to be included in the distribution requirements for all students, and equally acute in most science departments to design courses that will attract these students, regardless of whether such courses will have a lasting effect on them. Think for a moment if the science departments in our high schools had to justify their existence by serving only the 10 percent or so who *claim* to be science-bound, or if our college science departments served only its majors. Most science departments would collapse for lack of a critical mass of faculty, and the training of science professionals would have to be given over to a relatively few specialized state or national high schools and colleges. So, unseemly as it may be, a major factor in the pursuit of scientific literacy is self-justification and perpetuation of the science and science education professions. Many university science departments survive only by virtue of the "point credits" they earn through their introductory courses. How else to account for such promotional course titles as Physics for Poets, Kitchen Chemistry, Biology for Living, etc. Not that this is altogether improper, for it is essential that university science departments be able to support "critical masses" of faculty to carry on research and prepare future scientists. Whether this is equally true for all students at the high school level is open to question.

A second premise concerns general education in science. If the purpose of such education is to create a scientifically literate public, the principal target audience, namely the student population, may be ill-chosen. However one chooses to define scientific literacy, its main objective seems clear enough: society (and the individual) will somehow benefit if its members are sufficiently literate to participate intelligently in science-based societal issues. Assuming this is true, and on the surface it seems perfectly sensible, is the proper target audience the student or the adult population? Obviously, if the purpose of such literacy is to benefit society, it is really only the adult population that is in a position to contribute to the public good. So while students may attain some level of scientific literacy relating to the individual science courses they take in school, what good is it if they fail to retain this knowledge into adulthood?

The mistake we make is in assuming that because some, perhaps even many, of our students perform well in school science, they have achieved a measure of scientific literacy that will serve them as adults. Most science teachers leave their classes feeling that they have communicated successfully with some number of nonscience students, and most likely they have. But the end effect of this is a delusion. Good school performance, even a reasonable level of scientific literacy while one is a student, provides no assurance that the individual will retain enough science when he or she becomes a responsible adult, presumably contributing to the overall good of society. It would be different if science education for the general student were justified solely on the grounds of a cultural or intellectual imperative, but while many may believe that it should be, science educators in general fear that doing so would spell its doom by relegating it to purely elective status in the schools.

Here lies the crux of the matter. Whatever we may do to turn students on to science, to make them acutely aware of the world around them, and get them to at least appreciate what the scientific enterprise is about, if not so much science itself, we are guided in the schools by immediate feedback rather than long-term retention. After all, to be pragmatic about it, having literate students who turn out to be scientific illiterates as adults does not do much for society. We know that the staying power of science courses is very poor, but what is particularly depressing is the fact that although most students lapse back into scientific illiteracy soon after they graduate, they nevertheless think they are reasonably literate in science.

All surveys done on adult literacy in science demonstrate both their illiteracy and their complacent but misguided conviction of literacy. Fully 70 to 80 percent of U.S. adults surveyed in a poll several years ago considered themselves interested in scientific and technological matters, and concerned about government policy in science and technology, *and* rated their basic *understanding* of science and technology either as *very good* or *adequate*.[1] Similar surveys in Western Europe show the same results; the public generally believes it knows far more about science than it really does. A case in point is the "89 percent story." A recent Danish survey sought to determine how well informed the public was on biotechnology.[2] Eighty-nine percent of the respondents claimed they were either well or very well informed on the subject. In a contemporaneous survey in Ireland the question was put differently by asking the respondents to state in their own words what they *understood* biotechnology to be. Of these respondents, 89 percent professed ignorance, saying that they really knew nothing about the subject. In other words, when forced to prove their *understanding* of a particular part of science, one that is currently in the public eye, the public falls far short of acceptable norms. This huge disparity between what average adults actually know about science and what they believe they know strikes at the heart of the problem, for it means that most adults know all they need or want to know about science. Changing that perception is prerequisite to even thinking about ways to encourage adults that attaining scientific literacy may be worth the effort.

Such self-delusion is far more damaging to the well-being of society than professed ignorance of science. As the historian Daniel Boorstin put it so well, "The great obstacle to progress is not ignorance but the illusion of knowledge."[3]

▲ GENERAL EDUCATION AND SCIENTIFIC LITERACY

While some historians place the beginning of scientists' concerns with social issues in the Great Depression of the 1930s,[4] the active pursuit of universal scientific literacy is essentially a post–World War II phenomenon, having its origin in "civic" or "societal" concerns. Following the war, many scientists, having witnessed the horror brought on by one of their major scientific/military achievements (the atom bomb), believed that the best way to avoid such catastrophic use of science in the future was to educate the public to the potential of employing science for evil as well as for good, and to seek civilian control of nuclear energy. Thus was born a loosely structured but highly visible movement toward what came to be called "scientific literacy" for the general public. A number of scientist organizations and public interest groups were formed in 1945, notable among these being the Federation of American Scientists and the Federation of Atomic Scientists. Shortly afterward, in 1946, the Chicago chapter of the Federation of American Scientists began publishing the *Bulletin of Atomic Scientists* (later to become an independent journal), a highly regarded publication characterized by Daniel Greenberg as "the conscience journal of science."[5] The common element in all these activities was the belief that with the atom bomb, science and technology had brought civilization to the edge of a precipice, and that lacking a more reasoned decision-making process regarding the use of such destructive technologies, civilization might well push itself over the edge. As Robert Oppenheimer, the physicist who directed the development of the atomic bomb, said: "The physicists have known sin, and this is a knowledge which they cannot lose."[6]

Nuclear energy was only one of a number of social concerns perceived as having been created by science and technology. Today one hears more about such issues as (non-nuclear) pollution of the environment, toxic wastes, acid rain, nuclear power, depletion of natural resources, animal experimentation, genetic engineering, and the "greenhouse effect," to name but a few, all man-made and all seen in the minds of many as examples of the misuse of science. The obvious answer was education—of legislators through effective lobbying, and of society so it could exert its reasoned influence on the legislative process. Thus was born the notion that science literacy, as a major goal of science education, meant preparing students to cope intelligently with "science-based societal issues." To most, this meant that the individual be sufficiently sophisticated in science to reach *independent* judgments on those issues that are brought to public notice by special interest groups or through the mass media. That this may be an impossible task is rarely considered by those who are taken with the idea that responsible citizenship can only mean *personal* expertise in such matters rather than prudent reliance on the opinions of credible experts. We shall see that the scientific knowledge required for total dependence upon one's personal judgment in such matters is indeed impossible to achieve in the general population. If society really wishes to play an intelligent role in the resolution of science-based issues, other means than reliance upon formal science education will have to be found.

While educating the general public in science was not a new idea, the concerns brought on by the military use of science focused attention on a host of problems relating to the social and political dimensions of science and technology. Not only did the intensity of these postwar concerns far surpass all previous public interest in science, they also appeared to represent a marked departure from earlier rationales for general education in science, although this point may be debatable. In the prewar period, beginning in the second decade of the century, while some science educators, influenced largely by the philosopher/educator John Dewey, were discussing how to instill "scientific method" and to measure "scientific attitude" in students, Dewey also set the stage for "social literacy" as one of the purposes of science education:

> The business of the high school is primarily a social business, not of creating a class of specialists. The public that pays taxes for the support of schools is justified in requiring that whenever it can be accomplished without doing violence to the subject taught, the subject shall be so taught as to make individuals more intelligent and hence more competent in doing their share in social life. Contemporary civilization rests so largely upon applied science that no one can really understand it who does not grasp something of the scientific methods and results that underlie it; on the other hand, a consideration of scientific resources and achievements from the standpoint of their application to the control of industry,

transportation, communication, not only increases the future social efficiency of those instructed, but augments the immediate vital appeal and interest of the subject.[7]

Thus, while many of the science-based societal issues that face us today were not present in Dewey's time, he nevertheless believed that society would be better served if its educated individuals were reasonably schooled in science. Dewey's views on the social or "good citizenship" purpose of science education were never put to the test in an organized fashion, despite his enormous prestige in educational circles. Except for a few isolated experiments in bringing some social aspects of science into the science curriculum, it does not appear that a major move to design and implement science curricula with this objective in mind, and to evaluate the outcome, was ever made. Obviously, if it had, and had been proven successful, we would not now be groping for ways to achieve scientific literacy in the civic sense. Yes, practical physics and general science courses were in vogue for a time, but their design—how a telephone, or a steam engine or automobile, or machines generally work—could not possibly achieve Dewey's goal of a more intelligent citizenry. One may suspect, incidentally, that Dewey and his fellow educators of the time must have been faced with the same semantic problems confronting us today: what does it mean to be a "more intelligent individual," or to be "scientifically literate," and how should educators test for such qualities of the mind?

Dewey believed that high school science education should accomplish more than merely prepare the student in a practical way for eventually playing a useful role in society; he urged the development of what he called "scientific habits of the mind," which he believed could come from exposure to the methods of science:

> Scientific method in its largest sense is the justification on its intelligent side of science teaching, and the formation of scientific habits of the mind should be the primary aim of the science teacher in the high school. Scientific methods in their largest sense are more than matters of pure technique of measurement, manipulation and experimentation. There may come a period in the scientific training of scientific specialists when these things for the time being become ends in themselves. In secondary education their value and hence their limits are fixed by the extent to which they react to create and develop logical attitudes and habits of the mind. The methods of experimental inquiry and testing

which give intellectual integrity, sincerity and power in *all* fields, rather than those which are peculiar to his specialty, are what the high school teacher should bear in mind. A *new type of mind* is gradually developing under the influence of scientific methods.[8]

Dewey was not the first to see education in science as a means of developing desirable thinking habits. More than a decade earlier, Karl Pearson, probably best known for his work on statistical methods, in discussing the link between science and citizenship, extolled the virtues of reasoned thought:

> The classification of facts, the recognition of their sequence and relative significance is the function of science, and the habit of forming a judgment upon these facts unbiased by personal feeling is characteristic of what may be termed the scientific frame of mind. The scientific method of examining facts is not peculiar to one class of phenomena and to one class of workers; it is applicable to social as well as to physical problems, and we must carefully guard ourselves against supposing that the scientific frame of mind is a peculiarity of the professional scientist. . . . The importance of a just appreciation of scientific method is so great, that I think the state may be reasonably called upon to place instruction in pure science within reach of all its citizens. . . . The scientific habit of mind is one which may be acquired by all, and the readiest means of attaining to it ought to be placed within the reach of all.[9]

Developing such scientific habits of the mind appealed more to the innovative science educators of the 1920s and 1930s than did training students primarily for good citizenship. Many agreed with Dewey that getting students to think scientifically (i.e., logically) could be the single most important outcome of science education, but however much they tried then, and still do today, to instill rational thought into students, their efforts were, and still are, largely unsuccessful. Champagne and Klopfer concluded from a study of the literature since 1916 that despite the acceptance by many science educators of Dewey's belief in the value of scientific thinking to the student outside the field of science, they never learned how effectively to translate this doctrine into classroom practice.[10] Regardless of the failure of science educators to develop in their students Dewey's scientific habits of the mind, or reflective thinking, he never wavered in his conviction that this could be the most important outcome of general education in science. Writing a quarter-century after first stating his views on the purpose of science education for the mass of students, he said:

> [T]he responsibility of science cannot be fulfilled by methods that are chiefly concerned with self-perpetuation of specialized science to the neglect of influencing the much larger number to adopt into the very makeup of their minds those attitudes of open-mindedness, intellectual integrity, observation and interest in testing their opinions, that are characteristic of the scientific attitude. . . . Every course in every subject should have as its chief end the cultivation of these attitudes of mind.[11]

By this time, after intense effort by a small but growing community of science educators, science had become fairly well established in the high school curriculum. Dewey, however, became concerned by what he perceived as a growing trend toward overspecialization in science curricula, that is, by a tendency to design science courses mainly for those students showing an aptitude for science who might be counted upon to become professionals in the field. Dewey's concern that the high school science curricula were being aimed more at the self-perpetuation of scientists and science educators than at the general student is often voiced by educators today, but it is easy to see why Dewey would be alarmed by such a trend in his day. After all, like Pearson, Dewey believed that *all* students could benefit from some training in science by acquiring his hoped-for scientific habits of the mind, but this obviously could not occur if courses were designed primarily for would-be specialists and thereby attracted only a very small fraction of the student population.

There are many science educators today, this writer included, who would strongly agree with Dewey's philosophy on developing scientific habits of the mind, or scientific attitude—who would argue, in fact, that a society somehow cultivated in the discipline of rational or critical thought (often referred to in Dewey's time as "reflective thinking")[12] but knowing little about science proper would be better equipped to guide its affairs than one in which the individuals have a good grasp of factual science but know little about the methods and canons of science.

The "scientific attitude" referred to by Dewey is thought by some to have triggered the scientific literacy movement, or at least to have launched the initial efforts to assess in individuals something akin to what would later come to be known as scientific literacy. Jon Miller, for example, who has long been active in assessing scientific literacy in adults, dates the systematic study of the scientific literacy of the American public to the 1930s, citing attempts to define scientific attitude in behavioral terms and the development of tests to measure scientific attitude as evidence of such early interest.[13] As we shall see later, however, the concept of scientific attitude at the time was somewhat different from the earliest perception of scientific literacy. One definition of scientific attitude, for example, was proposed by a well-known University of Wisconsin science educator of the 1930s, I. C. Davis, based on a survey of several hundred experienced teachers throughout the country as part of an effort to develop a philosophy of science teaching:[14]

> We can say that an individual who has a scientific attitude will (1) show a willingness to change his opinion on the basis of new evidence; (2) will search for the whole truth without prejudice; (3) will have a concept of cause and effect relationships; (4) will make a habit of basing judgment on fact; and (5) will have the ability to distinguish between fact and theory.[15]

A *scientific* attitude exclusively? It might well serve as a judge's admonition to a jury, or the focus of a course in philosophy or logic, or a "how to" guide to a social order for the efficient conduct of its affairs. Clearly, the study of science is not the only route to such qualities, but it is a *possible* route, and were it not for the problems perceived by the average student as being inherent in the study of science, it could be an ideal means to this goal.

In any event, if a science curriculum is to develop such qualities, whether they be called scientific attitude, scientific habits of the mind, or critical thinking, one must have an accurate means of testing the end product. Everyone concerned with educational assessment would agree that this is the hard part; setting educational goals (essentially a "wish list") is easy by comparison. A number of so-called objective tests for scientific attitude were devised in the 1930s by Davis,[16] Hoff,[17] Noll (the "Wisconsin school"),[18] and others, but no reliable evidence was found of either a significant correlation with scientific *aptitude* or with school performance generally. This may be taken to mean that either the test items did not accurately examine the qualities characterized as scientific attitude, or what is more likely, assuming that they did reflect these qualities, students failed to relate the test questions to their perceptions of the nature of science. Thus did efforts to put Dewey's grand philosophy of science education into practice come to a disheartening end, although most science educators, even today, regard his "scientific habits of the mind" as one of the most constructive ideas in the history of science education. While his major objective of getting all students to think rationally was never realized, Dewey's legacy nevertheless persists; science is now required of virtually all students. It reaches down into the elementary grades and extends through high school and into the colleges. Whether it will ever succeed in developing scientific literacy is another question that we shall explore later.

▲   THE TURNING POINT

With the war, the decade of the 1940s was largely a holding period for innovation in science education, so it was not until the 1950s that active interest in the notion of science for citizens could be revived, and the drive for scientific literacy began gathering momentum. It was in the '50s and '60s that the concept of "scientific literacy" as a goal of science education really began to surface. Previously, the major goals of science education had a more practical bent, both in terms of curriculum and desired outcomes. The post–World War II period (the late '40s and early '50s) saw a rapid growth in American peacetime industry, with a corresponding need for many more scientists, engineers, and allied professionals, and obviously more science teachers. Hence the emphasis then was on getting more students into science-related careers.

As for those students not interested in scientific careers (the vast majority) some exposure to science had long been thought important, as we have seen, on the theory that a discipline so prominent in the affairs of a society should be part of the general education of all its students. But true scientific literacy, at least as most now view it, namely, *understanding*, among other things, the principal features of the scientific enterprise, was not the real objective. Instead, the goal was somehow equated with "science for effective citizenship," that is, developing an informed public capable of playing an intelligent role in science- or technology-based societal issues—not unlike Dewey's rationale for science in general education: to produce an educated public equipped to serve the needs of a democratic society. The term "scientific literacy" was never clearly defined in operational terms, seeming instead to appeal to many educators largely because of its vague promise of an intellectual Utopia. We have already seen, given the examples of Spencer and Huxley in the mid-nineteenth century, how thoroughly entranced one can become by visions of a scientifically literate society.

In 1954, the National Science Foundation (NSF), the independent federal agency whose principal function was (and still is) to support basic and applied research in science and engineering, began funding educational programs designed to increase the pool from which science and engineering professionals are drawn. Chief among these activities were a number of curriculum development projects intended to improve the quality and content of science and mathematics instruction at the secondary school level (the first being the well-known PSSC Physics curriculum), and teacher training institutes designed to upgrade the knowledge and skills of high school science teachers who would be teaching the new (and/or existing) curricula to their students were soon established.

Then, in October 1957, came the shock wave created by *Sputnik*, the first artificial satellite to orbit Earth, and the NSF education programs went into high gear. Determined not to allow the Soviet Union to surpass the United States in scientific and technological achievement, Congress increased NSF's education budget from three and a half million dollars to nineteen million dollars, and eventually to sixty-one million, and enlarged the agency's statutory authority, permitting it to support science, mathematics, and engineering education at all levels, including the elementary grades. At the same time, the National Defense Education Act (NDEA) of 1958 authorized the U.S. Office of Education, then part of the Department of Health, Education and Welfare, to make funds available to local school systems (through their state education departments) for some remodeling of facilities and for acquiring equipment and teaching aids to improve their teaching of science and mathematics. Huge sums (several billion dollars) went into these programs over the next two decades, more in current dollars than was spent on developing the atom bomb thirty-five years earlier. What had begun as an attempt to produce more trained scientists and engineers soon expanded, at least in the minds of many science educators, into an effort to provide all students, and the public generally, with a broader *understanding* of science and technology.

Science, as a result, became a major feature of the general pre-college curriculum in the United States, as it was in other highly industrialized nations. Where some state education departments failed to make science courses mandatory for high school graduation, colleges usually accomplished the same end by requiring a minimum science background for admission. And the colleges, in turn, perpetuated the notion that science should be part of the lore of educated adults by insisting that most graduates at least be exposed to it. At the outset there was great optimism, for it seemed unlikely that such a massive effort to reform science education could fail to trigger a marked improvement in what had come to be known as "scientific literacy," at least among the educated segment of the population. But fail it did. Notwithstanding the sincerity and effort of many teachers and administrators, these educational reforms proved largely ineffective. Granted, the public may be more sensitive today than it was forty to fifty years ago to some science-based issues—nuclear weapons, the war on cancer, pollution, waste management and other environmental issues, and so on—but its current *understanding* of the basic principles underlying such issues is no better now than it was just after the end of World War II.

The pace of curriculum reform slowed noticeably in the late 1970s, as questions were raised about the effectiveness of some of the programs that had been developed, and, as we have already seen, it virtually ground to a halt in the early '80s with the cutoff of funds to NSF's Education Directorate, which had come to be regarded as the mainstay of the science education enterprise in the United States. Following a brief hiatus, funding was restored in the mid-'80s and the NSF resumed its support of curriculum reform and related activities in science education. The first round of major curriculum reform in science education failed to achieve its hoped-for objectives, namely, attracting many more students into science-related careers (which was fortunate in a way because of the greater unemployment of scientists and engineers that otherwise might have resulted) and developing in the general public a reasonable level of scientific literacy, ill-defined though this latter objective may have been. It is questionable whether the current reform movement will be any more successful, even though one result has been action by many states to increase the amount of time for science and mathematics in their schools, and to increase as well the number of mandated courses in these subjects.

▲ THE MEANING OF SCIENTIFIC LITERACY

We have already seen that the concept of scientific literacy is clouded by vagueness and imprecision. It was a postwar movement that did not have the same meaning initially as one would normally associate with the term, namely, "understanding science." Instead, as John Burnham has pointed out, scientific literacy was first equated with science policy, this being at the time the primary point of intersection of science and society.[19] Had this meaning survived, the purpose of scientific literacy might have remained clear. But in the last half of the century, beginning with the post-*Sputnik* decade of the 1960s, the education community gradually broadened the term to mean the collective goals of science education for all, but especially for the general student, with the science/society interface being only one of the goals rather than the principal objective.

Yet many scholars, particularly social scientists but also some natural scientists, persisted in the view that scientific literacy meant primarily an ability to cope with the societal implications of science, from *understanding* what science does to exercising control over it. Benjamin Shen, a physicist-educator, distinguished three forms of scientific literacy—practical, cultural, and civic—referring to the last of these as the "cornerstone of informed public policy." "The aim of civic science literacy is to enable the citizen to become more aware of science and science-related issues so that he and his representatives can bring their common sense to bear upon these issues."[20] And the social scientist Kenneth Prewitt used the interesting phrase "scientifically savvy" to describe the individual who acts "on the basis of a shrewd understanding of the deeper principles and structures that govern complex situations," asserting that "from the perspective of democratic practice, the notion of scientific literacy does not start with science itself. Rather, it starts at the point of interaction between science and society. My understanding of the scientifically savvy citizen—in contrast to the scientifically savvy parent or consumer or employee or producer—is a person who understands how science and technology impinge upon public life."[21]

It was confusion over the differing connotations of scientific literacy, the attempt to subsume all under a common heading, from mastery of basic knowledge to national technological superiority, from science viewed as a cultural imperative to that of a social responsibility, from science content to science attitude, that finally led to serious questioning of its real purpose. The goals were simply too fragmented and not accepted with universal enthusiasm by the science education community. Yet despite the problems of definition, by the 1980s scientific literacy had become the catchword of the science education community and the centerpiece of virtually all commission reports deploring the supposed sad state of science education.

In all the confusion and rhetoric surrounding the rush to achieve this high-sounding ideal, too few recalled James Conant's earlier counsel that scientific literacy should mean mainly the ability to choose one's experts wisely. He wrote that a scientifically literate person was one who could "communicate intelligently with men who were advancing science and applying it."[22] Of all who entered the literacy arena, Conant appears to have been among a small number who concluded that making everyone expert enough to reach independent judgments on science-related issues was futile, and that a better approach would be to instruct them in how to obtain the best advice possible on such technical matters. Bear in mind that the goals of science education in the first half of the century were set largely by educators, that there was a brief period following World War II when scientists became involved in the process, but that partly by default this has given way again to control by the education community. It should also be noted that absent from all the rhetoric on the meaning of scientific literacy (much of it useful in setting the stage) was a recipe for achieving any of the stated goals.

▲   THE DIMENSIONS OF SCIENTIFIC LITERACY

All formal definitions of scientific literacy are bound by the conventional view that at the very least scientific literacy means having some grasp of science, the main difference being in the level of scientific knowledge that should be requisite for said label. At one extreme, some members of the scientific community would demand the equivalent of a baccalaureate degree, on the theory that to be considered literate in science one must have at least the knowledge expected of entry-level professionals in the field. At the other extreme is the view that little or no formal science is needed, that understanding the social problems brought on by science and technology is more important than knowing how to solve them, and that erring on the side of caution is the simple antidote for ignorance of science. Between the two extremes is a wide range of prescriptions for scientific literacy, none very specific in the sense of a curriculum that can be shown to achieve the purpose.

For the present, let us assume, as do most science educators, that scientific literacy requires some level of *understanding* of science. If so, attempts to fashion a simple definition of the phrase in absolute terms must fail because a meaningful definition can be expressed only in behavioral terms or measurable outcomes, that is, what should be expected of a science literate or one who "understands" something about science? Jon Miller, who has sought to establish useful criteria for testing scientific literacy, suggested at a symposium on measurements of Public Understanding of Science and Technology held during the 1989 Annual Meeting of the AAAS that "functional scientific literacy should be viewed as the level of understanding of science and technology needed to function minimally as citizens and consumers in our society."[23] Requiring that the individual be able to "function minimally" begs the question somewhat, for we have seen that many individuals actually believe they function quite well despite their poor level of scientific literacy. Miller fails to specify exactly what he means for one to function minimally, except to suggest that scientific literacy requires (1) a basic vocabulary, (2) an understanding of science process, and (3) an understanding of the impact of science and technology on society. There are many variants of this theme, but the same three basic ingredients are found in all; the only difference is in emphasis and the amount of detail included in each topic. Actually, one could argue that Miller's first two topics are really no different than those normally covered in conventional high school and college science courses. They are clearly among the earmarks of one who is competent in science, but it would be helpful to have a more explicit set of guidelines. Few educated individuals are totally illiterate in science; everyone knows some facts of nature and has some idea of what science is about, however naïve or misconceived these notions may be. Hence, it is an oversimplification to assume that an individual is either literate or illiterate in science. Instead, one might distinguish several levels of literacy, levels that are normally attained sequentially by science-bound students, for example, during their formal exposure to science. Following are descriptions (definitions, if you will) of three such levels of literacy, which build upon one another in degree of sophistication as well as in the chronological development of the science-oriented mind, and which, because of their vertical structure, may be useful as criteria for judging scientific literacy, if a knowledge of science is indeed to be included in such literacy.[24]

    1. *Cultural Scientific Literacy* Clearly the simplest form of literacy is that proposed several years ago by Edward Hirsch in his highly popular book.[25] Hirsch argues that "cultural literacy," by which he means a grasp of certain background information that communicators must assume their audiences already have, is the hidden key to effective education. He goes on to provide a list of several thousand names, dates,

places, events, and so on illustrating the type and scope of knowledge that is shared by literate Americans. Included are several hundred science-related terms which, with appropriate definitions, are claimed to constitute a lexicon for the scientific literate—obviously a necessary but hardly a sufficient condition for scientific literacy beyond its most primitive meaning. Note the similarity between this and the basic vocabulary cited by Miller. Note too that if all one needed were such a lexicon, scientific literacy would be easy to acquire by rote; a companion dictionary, incidentally, was subsequently developed by E. D. Hirsch, J. F. Kett, and J. Trefil.[26] Yet this is the only level of literacy held by most of the educated adults who believe they are reasonably literate in science. They recognize many of the science-based terms (the jargon) used by the media, which is generally their only exposure to science, and such recognition probably provides some measure of comfort that they are not totally illiterate in science. But for the most part, this is where their knowledge of science ends.

2. *Functional Scientific Literacy.* Here we add some substance to the bare skeleton of cultural literacy by requiring that the individual not only have command of a science lexicon, but also be able to converse, read, and write coherently, using such science terms in perhaps a nontechnical but nevertheless meaningful context. This means using the terms correctly, for example, knowing what might be called "some of the simple, everyday facts of nature," such as having some knowledge of our solar system, of how the Earth revolves about the Sun and the Moon revolves about the Earth, and how eclipses occur. Much of this is perhaps better classified as natural history than as science proper, but it is nevertheless a part of the overall scientific enterprise. Elementary, to be sure, yet unfortunately not at the fingertips of most Americans. Or to get a bit more sophisticated, expecting the individual to identify the ultimate source of our energy; or the "greenhouse effect," or knowing what "clean air" means or how we get the oxygen we breathe. And getting still more sophisticated, hoping that the individual knows the difference between electrons and atoms, or what DNA is and the role it plays in living things.

One could go on and on listing such basically simple facts about nature, and most objective tests of literacy are based upon just this kind of knowledge or recall. Not a very demanding requirement, yet estimates of the number of adults in the United States, for example, who might qualify at this functional level are distressingly low. Miller, in the paper cited above, places the fraction of adult Americans who possess a "minimal" understanding of scientific terms and concepts at about 30 percent. This number seems very high, but it is clearly a soft number, depending as it does on the sophistication of the test items designed to assess *minimal* understanding; obviously one could easily halve this figure by a more rigorous selection. Nonetheless, whatever the number, it does indicate that the science recall of most adults is very poor, since most of the terms and concepts tested probably would have been at their fingertips during their school years. It also shows that whatever exposure they may subsequently have to science, mainly through the mass media, fails to reinforce such basic information about the natural world.

A useful distinction between these two levels of literacy is that the *functional* literate should be able to engage in a meaningful discourse on most science articles that appear in the popular press, at least posing intelligent questions, where the *cultural* literate is in the position of one who knows a smattering of a foreign language but can only stare blankly when one fluent in the language seeks to initiate a flowing conversation. Determining whether an individual qualifies for one or the other of these forms of literacy (functional literacy presupposes cultural literacy, of course) through objective tests is relatively easy, since both depend largely upon factual recall. But this is not what science is really about. What is lacking in both these levels of literacy are (a) the process of science, that is, a prescription for seeking out and organizing the factual information in the unique manner that is characteristic of science, and (b) the fundamental role played by theory in the practice of science. Incorporating these brings us to the ultimate (true) level of literacy, the kind most difficult to attain by general students and to evaluate objectively by teachers:

3. *"True" Scientific Literacy*. At this level the individual actually knows something about the overall scientific enterprise. He or she is aware of some of the major conceptual schemes (the theories) that form the foundations of science, how they were arrived at, and why they are widely accepted, how science achieves order out of a random universe, and the role of experiment in science. This individual also appreciates the elements of scientific investigation, the importance of proper questioning, of analytical and deductive reasoning, of logical thought processes, and of reliance upon objective evidence. These are the same mental qualities that John Dewey called "scientific habits of the mind" nearly a century ago and which he proposed as the main rationale for compulsory science education, a rationale that today is

often called "critical thinking." Whatever term one chooses, however, it remains a mental state that has never come to pass in the general population. This is obviously a demanding definition of scientific literacy and some may argue that it is designed to make such literacy unattainable by the public at large. But it means only that the term itself, "scientific literacy," has been used too loosely in the past and that, when viewed realistically, true scientific literacy, as defined here, is unlikely to be achieved in the foreseeable future. Yet some important elements of it may be. Notice that we do not require our scientific literate to have at his or her fingertips a wealth of facts, laws, or theories (the major conceptual schemes in science can virtually be counted on the fingers of both hands), or be able to solve quantitative problems in science. Nor are advanced mathematical skills essential; we should only expect that our literate understands and appreciates the central role played by mathematics in science. Nevertheless, for reasons that will be discussed later, even this modest criterion puts scientific literacy beyond the reach of most educated individuals. This writer's estimate of the fraction of Americans who might qualify as true scientific literates in this sense is 4 to 5 percent of the adult population, nearly all being professional scientists or engineers. Miller places this number at about 6 percent by his benchmarks. It is worth noting that these low estimates of scientific literacy in the adult population are not confined to the United States; a companion study to Miller's that was conducted in the United Kingdom produced a 7 percent estimate.

The first level is met by almost all high school graduates, and the second by those who might be classified as "serious" students, say in the upper quartile of their classes. But if not science-bound, most apparently fail to retain such knowledge into adulthood. One would expect all science-bound students to achieve the third level, as far as science is concerned, by the time they graduate college, if not sooner, but whether they can extrapolate their skills to societal issues remains questionable. As for the non-science-bound, very few ever achieve it unless they are very strongly motivated—and well taught.

▲ FAILURE OF THE SCIENTIFIC LITERACY MOVEMENT

If "true scientific literacy," or some meaningful version thereof, could be achieved in a reasonable fraction of the adult populations of our highly industrialized nations, say 20 percent rather than the current 5 percent, we should have much less concern about the public image of science and technology, or of society exercising sensible controls over these enterprises. With this core of knowledgeable adults participating in the democratic process we could have some confidence in the outcome. This is not to suggest that the problem would then be fully resolved, for so many of the "science-based societal issues" are also emotionally burdened, which often moves them out of the arena of reasoned debate; but at least any discussion of the science or technology involved should be better informed. The other two levels of literacy described above would not achieve the same purpose, for in order to diminish public concern about science and technology, at least the educated public must learn to appreciate what these enterprises can do and how they work. We have already made the point of stressing adult literacy in science, rather than merely student literacy, for the latter is no guarantee of literacy in adulthood, which is essential if we are to have an enlightened society. Twenty percent happens to be the percentage of college graduates in our adult population, which suggests that if all college graduates were literate in science we might have a markedly different society. Such a "20 percent solution" is discussed more fully in a later chapter.

The need for widespread scientific literacy has been rationalized on several grounds, but when carefully probed most have a hollow ring to them. High on the list, as we have seen, is the promotion of good citizenship by enabling the individual to act on societal issues having a scientific base. In theory, an informed electorate, meaning in this context one that is scientifically literate, should be better able to deal with such issues, to reach independent judgments, and to elect government officials who properly reflect these judgments.

We have already seen that few such issues are truly based in science, as distinguished from technology, but some are, wholly or partly, based in science, and it may be instructive to look at a few examples. Funding questions in the area of "big science" certainly qualify as science-based issues affecting American society simply because budgets in support of science and technology are limited and choices must be made, ultimately by Congress and the people. A few years ago, when the Soviet Union still existed and the Cold War was still warm, the Strategic Defense Initiative (SDI) or "Star Wars," as it was commonly called, was being stubbornly advanced by the administration as a first line of defense against long-range ballistic missiles. Opponents of the proposal argued on two separate grounds. One was the political wisdom of inviting possible retaliation by developing a weapons system that might be viewed as having offensive capabilities rather than only the defensive strategy its name implied. The other was technical: what were the prospects of developing such a weapon, based upon yet-to-be-developed high-power X-ray lasers, to be 100 percent effective against a well-organized missile attack, and assuming that the laser problems could be worked out, how would one test the "umbrella," other than by computer simulations, against an anticipated five to ten thousand enemy missiles, some active, others dummies, coming over within a short time? Only a few need get through to defeat the system. Politics aside, this was a reasonable question in light of the huge R&D costs projected for the system. Professional scientists lined up on both sides of the issue, some contending that it was technically feasible to achieve the goal, but most arguing strongly that apart from any moral considerations the project was impossible within the projected cost and time constraints.[27] How could the average citizen, even one truly literate in science, reach an independent judgment on this issue if professional scientists disagreed on the outcome?

SDI was a scientific issue that caught the public eye mainly because of its military overtones, but also at least partly for its budgetary implications. There are others that qualify for public attention largely because of substantial funding requirements. Two such major projects were strongly supported by the administration in their early stages. Both involved pure science but only one promised some early *practical* return. One was the huge accelerator, the SSC (Superconducting Supercollider); the other the Human Genome Project, on which some work was underway but which still required a long-term financial commitment. The SSC was estimated to cost at least eight billion dollars over ten years, the genome project perhaps three billion over a fifteen-year span. Assume that a decision had to be made between these on budgetary grounds. Should an informed electorate choose the less costly one, or the one offering the greater scientific merit? Surely the latter criterion ought to be foremost in our minds if the purpose of supporting scientific research is to improve our basic knowledge and hence the future quality of life. But this poses a major dilemma: how to evaluate their relative importance, both in terms of new knowledge to be gained and the probability of practical applications of such knowledge. The genome project would unravel the DNA in the human chromosomes and map the 100,000 or so genes contained therein, thereby revealing the chemical "instructions" for creating an individual at the molecular level, including the possible identification of one's physical and behavioral characteristics and one's predisposition to disease. It is easy to predict that with such knowledge the practical results in terms of improved health care alone might be substantial, but even this argument is questionable in the minds of some researchers. What about the proposed accelerator, which promised only to permit science to see deeper into the structure of the fundamental particles of nature, a "voyage of discovery," as some would call it?

The choice seems simple, does it not? The Human Genome Project would cost less and promises practical results. But can one really predict what the SSC might bring in the way of a comprehensive theory of the origin of the universe, or even in terms of unforeseen yet immensely valuable practical results? Can anyone truly forecast the possible fallout from basic science? Obviously, the choice is not that easy, except possibly to politicians who are motivated more by parochial than by scientific considerations. Even the scientific community would be hard pressed to choose between the two on scientific grounds alone and would be more likely to vote their parochial interests, with most physicists favoring the SSC and life scientists the genome project.

By the start of the Clinton administration in 1993 both projects had managed to survive the ax of budget-minded officials, but not without bruises from covering some rocky political ground. However, a year later the SSC, whose projected budget had grown some twofold since its inception, came under heavy fire from Congress and was finally closed down, wasting several billion dollars that had already gone into the program. More than the financial waste was the disruption of the professional lives and careers of hundreds of high-energy physicists. Obviously, this was not a scientific decision but one based on social and political (and hence financial) considerations. The story of the demise of the SSC is yet to be fully documented, but when it does appear there is likely to be a great deal of finger-pointing and deploring of the scientific shortsightedness of our politicians. Fortunately, the Human Genome Project is still alive and reasonably healthy, although not without its detractors—scientific, social, and political.

Two other large-scale projects currently in their preliminary stages are also worth noting because of the huge costs involved and the wariness of Congress to continue uncritical funding of such projects. Both are NASA projects: one, the *Freedom* space station, estimated to cost some thirty-seven billion dollars by its completion in the late 1990s; the other the Earth Observing System, a thirty-billion-dollar array of six large satellites to be orbited in 1997–98. A primary purpose of the space station would be to provide a gravity-free environment in which to study the behavior of matter and possibly develop new technologies, while the Earth Observing System would be an environmental research tool aimed at learning how the global environment works and is modified by human activities. Both have been criticized by some experts as being too grandiose, and by others as not worth the investment in terms of expected practical rewards. But both also have strong support among many scientists and government officials. Another objection voiced by some in the scientific community against *all* big science projects is that they tend to squeeze out "little science" in the competition for financial support, meaning that the average university-based scientist finds it more difficult to obtain research grants and support for graduate students. Who should make the final choice? In the face of all the pros and cons expressed by scientific experts regarding the priorities that should be assigned to such projects—often subjective, to be sure, but just as often free of bias—how should the lay public be expected to reach considered opinions on its own, assuming it really wished to do so?

Unfortunately, for reasons that in part have been touched on here and elaborated elsewhere,[28] I believe there is not the slightest possibility of achieving the level of scientific literacy needed to make such decisions independently in an appreciable fraction of the adult population. Several reasons can be offered to account for our failure to achieve anything bordering on scientific literacy, but the basic reason is that becoming and remaining reasonably literate in science requires a special effort on the part of students, a commitment that very few non-science students are prepared to make. Science is not easy—there is no getting around this—and the more we try to simplify it the more we find ourselves moving away from those areas that comprise the essence of true scientific literacy.

As far as the noncommitted student is concerned, science is difficult because of (a) its cumulative nature, which makes it necessary to build one's understanding of it like a tall edifice, layer by layer; (b) its repeated failure to accord with "common sense," giving it a somewhat mystical quality; and (c) its mathematical component. Indeed, we have seen that in the view of many scientists and educators, it is its cumulative property that most distinguishes science from other forms of intellectual activity. As for common sense, we know that while scientific inquiry generally begins with observations in the real, or everyday world, and in the end returns to that world in the form of technology (i.e., in the form of practical things), the steps in between, where the real science is done, are largely unfathomable to all but specialists. We have no commonsense counterparts for photons, genes, cells, novas, or black holes. When scientists talk about these phenomena they reason by means of models or abstractions, inventions that may seem reasonable to the scientist but are more like "black magic" to the lay person. Thus, taken together, the cumulative nature of science and its reliance on descriptions that often defy common sense make it necessary to devote extraordinary effort to the task of becoming scientifically literate. And we have yet said nothing yet about mathematics, the quintessential language of science, which so many students feel must have been invented to further complicate their studies rather than to make science more tractable. We are not a quantitative society, and mathematical illiteracy, or innumeracy, as it is now fashionably called, is as much a problem on its home ground as it is in science.[29]

▲ SCIENCE IS HARD

It is often said that were it not for mathematics, science would be relatively easy and scientific literacy would be commonplace. Indeed, one can always find some science courses, often elective college courses, that require no mathematics, either as prerequisites or as part of the course content. Their purpose, of course, is to attract the mathematics-shy student. Can science be taught without mathematics? Indeed, a good part of it can be, but the real power of science lies so much in its theoretical structure, which in turn usually relies heavily upon mathematical representation, that avoiding the math means leaving a substantial gap in one's appreciation of the nature of science.

Biology and to some extent earth science, being more descriptive than either chemistry or physics and involving much less in the way of mathematical requirements, have traditionally been the courses of choice for students wishing to satisfy a science requirement in high school or college with the least risk to their grade average. One can always present the purely descriptive aspects of a science without mathematics, which in the past has led some biologists to contend that scientific literacy can be attained with little or no competency in math, namely, through the study of biology. It is true that some of the earlier conceptual schemes in biology, such as the theory of evolution or the gene theory of heredity, can be explained with little recourse to mathematics, and most biology textbooks would bear this out. But as molecular biology gained prominence during the past few decades, modern biology became more firmly grounded in theory, particularly in molecular structure and the cross-disciplines of biochemistry and biophysics; and the future is bound to see both biology and earth science much more dependent upon mathematics than in the past, not merely in research but in teaching as well. In general, as a science matures, it passes first through a purely descriptive stage, then proceeds to an experimental stage, and finally, as meaningful patterns are seen to emerge, to a theoretical stage, where one usually finds it highly productive to use the language of science (mathematics) to describe natural phenomena and uncover new knowledge. Physics and chemistry are already at this final stage; biology and earth science, involving as they do far more complex systems, have been slower to mature to this level. So while it may be possible now to teach these disciplines with little or no mathematics, this will certainly not be the case in the future.

It is unrealistic to believe that one can fully appreciate the broad reach of science without seeing firsthand the role played in it by mathematical reasoning. Thus, innumeracy and scientific illiteracy remain closely linked and should be regarded as two facets of the same problem. Interestingly, a recent study by the Higher Education Research Institute (HERI) shows, perhaps not surprisingly, that the average attrition among college students who chose science, math, or engineering as their concentrations when freshmen was 40 percent by the time they became seniors. What is surprising is that the decline in the physical sciences was only 20 percent, while in the biological sciences it was 50 percent. Not surprising was the finding that the better the student's precollege education in science, the less the likelihood that the student will drop out of science.[30]

I have not said much about the special mode of thought that is said to be characteristic of science. As pointed out earlier, the habit of rational thought, or logical reasoning, or critical thinking, or "scientific habits of the mind," as John Dewey called it, is not indigenous to our advanced societies and never was. Hence students discover still another source of discomfort in science: their fantasies, speculations, and ingrained misconceptions about the workings of the universe must now be replaced by factual observations, logical thinking, and reliance on sound evidence, a transition that most students find difficult and are unwilling to make. By the time they reach secondary school, most students also find that their previous exposure to the purely descriptive aspects of science—the so-called wonders of nature but perhaps better-characterized as "natural history"—that they once found so simple and even exciting had become lifeless and deadening. Facts, definitions, formulas, names, dates, taxonomy—is this what science is really about? Of course, everyone must know some facts of nature, in the same sense that one needs to know various facts of history, or geography, or literature, but it comes as a rude awakening to students that far more important is an *awareness* of *how* such facts, definitions, classification schemes, and the like are determined, and how they contribute to the development of the conceptual framework of science.

▲   SCIENTIFIC LITERACY: SOME DISINCENTIVES

Coupled with the difficulty of learning science is the fact that there is no real incentive for students (or adults) to make the commitment necessary to become literate in science. Most individuals, certainly the vast majority of students who do not plan to become scientists, appear to be unwilling to make this effort. This should not be surprising, for after all, what incentive do they have? As the sociologist David Harman points out, adult illiterates learn to read and write when they realize that being literate is in their own self-interest.[31] Yet it is estimated that 20 percent of America's eighteen-year olds are functionally illiterate, let alone scientifically illiterate, a higher percentage than is found in any other highly industrialized nation.[32] The same is certainly true of science; students and adults will evidence a genuine interest in scientific literacy only when they are convinced that becoming literate in science is for their personal good, not simply because science educators or commission reports tell them so. If we are unable to convince some 20 percent of our young work force that ordinary functional literacy is essential to their lives, how can we persuade the bulk of our students that scientific literacy is important? Nonscience students want to know why they should be forced to study science if they have no interest in the subject.

The popular argument generally offered is that it better prepares a nonscientist to function in business or professional life. If there is any truth to this, students fail to perceive it, and small wonder: they need only look at their own professional family members and friends, at wealthy businessmen and powerful public officials, at people in the arts and entertainment and professors of humanities—all successful and respected members of society and most, if not all, illiterate in science. After all, what bearing does a lawyer's *understanding* of the double helix have on the success of his practice? How many times does a banker call on the uncertainty principle to make an investment decision? Is it necessary for the mayor of New York (or any other city not earthquake-prone) to be versed in plate tectonics to run City Hall, or for a surgeon doing laser surgery to understand the physics of lasers? Would a knowledge of chaos theory have boosted the careers of Luciano Pavarotti or Laurence Olivier? The same question might be asked of all educated adults in the work force, with essentially the same answer: there is no convincing evidence that *understanding* science is important to them. While enjoying the everyday comforts and benefits derived from science and technology, society has managed to insulate itself from any actual or even perceived need to understand their origins. The sad but simple fact is that one does not need to be literate in science (or mathematics, for that matter) to be successful in most enterprises or to lead the "good life" generally. It is sometimes said that all the science or mathematics that the average person really needs to know to get along in life can be written in the space of a matchbook cover. Hyperbole? Perhaps, but unfortunately not far from the truth.

Another line of argument often given students relates to their future role in America's technology-oriented economy. What will the job market for nonprofessionals be like? We have seen earlier that there is a noticeable trend away from science and engineering careers into other, more satisfying and lucrative fields. Will office workers who can handle computers be in greater demand than those who cannot? Will manufacturers be looking for factory workers who can operate the robots that will displace much of our labor force in the future? The answer to both must be "yes," of course; but does this mean that general literacy in science will be required? Probably not, as the history of technology clearly shows.

Manufacturers fully realize that their customers cannot be expected to understand the inner workings of their products, or even to handle them with great care. Hence they try to design their equipment to be as "idiot-proof" and durable as possible within economic constraints, and the consumer expects this of them. Simple instruction manuals, and occasionally a demonstration, are all that the average consumer needs to operate what is often highly sophisticated equipment. And the manufacturers provide warranties and service to restore the equipment to operating condition in the event of failure. Suppliers could not remain in business very long if they did not make it as easy as possible for the consumer to use their products. Moreover, the public has also come to rely on various government agencies and consumer groups to protect it from unsafe or hazardous equipment, foods, health practices, quack remedies, and so forth. Not that a knowledge of science wouldn't be helpful in some cases for diagnostic purposes and simple repairs, or perhaps more importantly, to avoid being taken in by exorbitant claims, but it is difficult to rationalize scientific literacy on such a slender basis. And the public certainly does not see it this way.

Some level of technological literacy might be useful, but even this, as a general rule, can be challenged. We learned to use electronic typewriters, office copiers, automobiles, FAX machines, video equipment, electronic machine tools, and the like through specific on-the-job training or simple written instructions, without knowledge of their inner workings, because such devices are specifically designed for use by nonexperts. So it is somewhat specious to contend that achieving scientific (or even technological) literacy will better prepare students for the nonprofessional job market, and students seem to realize this.

If, indeed, as much as one-third of our total population is functionally illiterate or worse, and, as we have seen, if well over 90 percent is scientifically illiterate, why is there so little organized national clamor over this state of affairs? The most obvious answer is that the problem has yet to be acknowledged *officially* at the highest levels of government, and this is unlikely to occur until most citizens are willing to accept the fact that we have some serious deficiencies in our social and educational goals. There is yet another reason for the failure of scientific literacy, one that normally is not cited. As will be shown later, the movement lacks the *unqualified* support of large sectors of the academic community.

The reader may have noticed that in this chapter a deliberate attempt has been made to avoid using the term "understanding" as a synonym for scientific literacy. Instead, except where the term appears by reference to its use by others, or in a direct quote, it has been accented to draw attention to the fact that the term poses almost as much of a semantic problem as does "scientific literacy" itself. In discussing Jon Miller's criteria for scientific literacy, we pointed out that requiring one to "understand" a given concept in science only begs the question, for this term also demands careful definition. If one means to *understand* in behavioral terms, that is, to act in accordance with such understanding, then the desired behavioral objectives must also be specified. This is rarely done, of course; most of us use the term as though it carries the same meaning to everyone. Yet without some objective measures it can be as vague a term as scientific literacy.

It would be best, of course, to avoid using the term in relation to scientific literacy, but since that is impractical one should be particularly sensitive to what is intended by *understanding* and qualify it whenever possible. A useful view of what should be implied by "understanding" or "knowing" something in science parallels the early constructivist concept of knowing as being able to explain it satisfactorily to others. I would subscribe to this meaning of the term, with the qualification that the "others" must include those already knowledgeable in the subject area (e.g., science teachers), as well as those who are ignorant of the subject. In other words, a sufficient condition for *understanding* is when the learner can in turn "teach" what he or she has been taught. Thus, to *understand* means to be able to explain, not simply by rote, but with the ability to extrapolate to other related examples. For example, we explain the fall of an object to Earth by invoking a gravitational force between them. The individual who *understands* this should also be able to *explain* how the Earth "falls" into an orbit about the Sun, or a satellite into an orbit about the Earth. Otherwise, the individual does not truly understand the concept of universal gravitation as postulated by Newton.

CHAPTER 5

The "Two Cultures"— and a Third

The great law of culture is: let each become all that he was created capable of being.
—Thomas Carlyle (*Richter,* 1827)

▲   No major educational movement has much chance of succeeding without the support of various constituencies, not the least being those that hold in their hands the intellectual and cultural mores of a society. Among the many problems confronting any attempt to achieve scientific literacy in the general public, or even in students, for that matter, is the lack of enthusiasm for such literacy exhibited by many university intellectuals who serve as academic role models.

Every generation not only has its share of new intellectual problems but often must deal with some inherited from earlier generations—problems that seem destined never to be fully resolved but to continue confronting future generations. One such issue, the "two cultures" controversy, a debate on the relative merits of a strictly literary versus a scientific education, first surfaced seriously in the mid-nineteenth century, then simmered beneath the surface awhile, only to emerge again in the middle of the present century, when it created a brief but stimulating diversion from the usual intellectual discourse. Two fairly recent best-sellers, Allan Bloom's *The Closing of the American Mind*[1] and E. D. Hirsch's *Cultural Literacy,*[2] both critical of perceived failings in American education and coming at a time of renewed efforts toward achieving the ill-defined and thus far elusive goal of scientific literacy for the masses, bring to mind the controversy of a generation ago provoked by C. P. Snow's spirited and widely quoted essay on "The Two Cultures."[3]

Not that Bloom and Hirsch, nonscientists both, deliberately set out to fan the dying embers of this old controversy. Yet the end result promised to be much the same. Bloom trivialized C. P. Snow's thesis as "silly conceits," and Hirsch treated the problem of literacy in science in much the same way that he did for the humanities and the social sciences, namely, by reducing it to one of communication via an agreed-upon lexicon (a necessary but hardly a sufficient condition for scientific literacy). Both have the typical educated person's view of science: an enterprise that has an enormous practical impact on our daily lives but somehow lies outside the intellectual mainstream. They realize that all efforts to bring science into our common culture, that is, to develop some degree of scientific literacy in our educated adult population, have failed. But where Bloom shrugged this off as unimportant, citing Max Weber's truism that reason (or science) cannot establish human values, Hirsch at least sought a painless shortcut to scientific literacy.

It is easy to predict that Hirsch's shortcut will not work; more difficult to comprehend is Bloom's obvious predilection for the past, which colored his view of all modern academic disciplines, except perhaps natural science. He put science off in a corner by itself, self-sufficient and successful and doing useful things, but whether science ought to intrude upon the intellect of the educated person appears to be a matter of indifference to Bloom. He is not necessarily wrong in this. But the fact that his and Hirsch's views are likely to impress still another generation of students cries out for some balance, which is perhaps best achieved by reviewing this most recent of the "two cultures" controversies, and the history behind it. Let us hope that enough time has passed since Lord Snow galvanized the intellectual community some thirty years ago to a critical self-appraisal that one can look dispassionately at the controversy and possibly learn something useful from the argument. This was impossible to do at the time, when tempers flared and the protagonists on both sides were urged on by partisan reinforcements. Not that there is much more hope now of resolving the issue, if indeed there is one to be resolved, but at least such a review may help new generations of students to appreciate the background of the science/humanities controversy and to understand in part why scientific illiteracy is so endemic to our society.

▲ BRONOWSKI AND C. P. SNOW

In the midst of the post–World War II reformation in science education, and shortly before Sputnik, the American educational scene was illuminated by two English scholars who sought to recast science in the human drama, to show that the modern scientific revolution, instead of dividing our culture, actually ought to unite it. One was Jacob Bronowski, mathematician and poet, subsequently to become well known as host/narrator of the popular BBC television series, "Ascent of Man." The other was Sir Charles Snow, scientist and novelist. Both were warmly received by the science education community, which understandably found their views on scientific literacy strongly supportive of the renewed thrust in science education. Yet these scientists *cum* humanists had little visible effect on the activities of science educators, who continued to experiment with the new science curricula, new theories of learning, and new models of in-service teacher training that characterized the free-spending period of the 1960s and 1970s in science education. In the end, as had their predecessors, both failed in their mission, which was to promote public concern for and understanding of science.

Bronowski, in a 1956 essay, took as his theme "The Educated Man in 1984,"[4] a date made famous by the frightening predictions of George Orwell's well-known novel.[5] After warning of the danger in having only relatively few members of a democracy versed in science, Bronowski predicted and cautioned further:

> It is certain that the educated man in 1984 will speak the language of science. This is not at issue. The issue is something else. Will the educated man in 1984 be a specialist, a scientist or technician with no other interests, who will run his fellowmen by the mean and brutal processes of efficiency of George Orwell's book? Or will he be a statesman, an administrator, a humanist who is at home in the methods of science, but does not regard them as mere tools to efficiency? The choice between 1984 and an earthly paradise does not depend on the scientists but on the people for whom they work. And we are all the people for whom science works.

And, in his effort to alert the public to the perceived danger of a culturally divided society, Bronowski also invoked images of Aldous Huxley's 1932 classic, *Brave New World:*[6]

> A world run by specialists for the ignorant is, and will be, a slave world. A man of taste who sneers at machines, a housemaster with his eye on the preponderance of university scholarships in classics, a civil servant who still affects to despise science, is abdicating his share of the future and walking with open eyes

toward slavery. By leaving science to be the vocation of specialists, they are betraying democracy so that it must shrink to what it
became in the decline of Athens, when a minority of educated
men (who had to be paid to make a quorum) governed 300,000
slaves. There is only one way to head off such a disaster, and that
is to make the educated man universal in 1984. This is the force of
my argument here, to make the language of science part of the
education, the cultural education, of the young who will have either to make or to suffer 1984.

Persuasive prose that hit the mark, particularly with social scientists,
but history has shown Bronowski's forecast to be little more than wishful thinking. His prophecy was wrong; the "educated man" in 1984 did
not speak the language of science—and still does not. Nor do our nonscientists generally, be they statesmen, administrators, or humanists,
feel at home in the methods of science. And least of all do our scientists
"run their fellowmen." To the contrary, a complaint frequently voiced
by the scientific community is that to a large degree the course of
science, through its financing, is controlled by individuals outside
of science. Yes, science is, and has long been, part of the education of
the young, but it has failed to penetrate their culture. Bronowski's forecast of doom has not come to pass. Nor is there any credible evidence
that the scientific community has any desire to blaze the path for our
society; indeed, were they to seek this role the counterbalancing forces
in a democratic society would quickly put an end to it. True, science
benefits all of us in large measure, but as we have seen, by and large the
"educated man" finds that he can reap these benefits with little or no
understanding of science.

    One might turn Bronowski's argument about, in fact, by suggest-
ing that it is just this separation of cultures, like the separation of
powers in a democratic form of government, that prevents one from
overpowering the other intellectually and completely winning the
minds and hearts of the public. True, the humanities and social sciences, in terms of numbers at least, and their influence on the lives of
students, dominate life on almost every university campus, but while
this may seem incongruous to some academic scientists, such domina-
tion is understandable when viewed in light of the kinds of activities
that engage most of society.

    Where Bronowski argued eloquently, albeit ineffectively, for sci-
ence understanding by all citizens, C. P. Snow sought the same goal
through a different route; but in doing so he managed only to revive a

century-old controversy. In a 1956 *New Statesman* article, later elaborated in a Rede lecture, Snow deplored the lack of communication between members of what he christened "the two cultures," referring to the two groups of intellectuals—scientists and humanists—that one finds on every university campus. But while his argument focused mainly on the academic community, which had been his chief influence, he generalized it to include the intellectual life of all of Western society, asserting that this society "is increasingly being split into two polar groups . . . at one pole we have the literary intellectuals . . . at the other scientists." To make his point, he posed, as a test of scientific literacy for humanists, the question: "What do you know of the Second Law of Thermodynamics?," equating this to what he considered a corresponding test of cultural literacy: "Have you read a work of Shakespeare?"[7] Several years later, in a "second look" at the two cultures, Snow acknowledged that he might have chosen a better example, a subject that deserved to be at the forefront of our modern-day common culture, but one that is not as demanding of background knowledge as was a true understanding of the Second Law; namely, a question such as: "What do you know of Molecular Biology?"[8]

Not that Snow came to feel that the Second Law had become less important in the scientific world; on the contrary, he realized that because of its great depth and generality, simply being aware of the existence of such a tenet of faith was of limited import, and that it was perhaps unreasonable to expect the nonscientist to have a meaningful understanding of it without first having a broad exposure to science. There is no way around this, and in a real sense it goes to the heart of Snow's argument that there exists an imbalance between the two cultures, literary and scientific, that tilts in favor of the literary culture. Reading Shakespeare requires at a *minimum* that one understand the common language in which it is written. Reading science requires this and more; one also needs the additional vocabulary of science, often the special language of science (mathematics), and particularly in the case of the Second Law, a reasonable grasp of the process and structure of physics. Since thermodynamics was then, and remains today, outside the common twentieth-century culture, although some might argue that it should not be, to press his point Snow sought another branch of science which, by most standards, would be requisite knowledge in the common culture, one that is reasonably self-contained and, unlike thermodynamics, does not involve great conceptual difficulties.

It is interesting that he chose molecular biology as a suitable alternative, for in the little more than a quarter century since then this branch of science has figuratively exploded, both in the laboratory and in the public press. Yet, despite its great scientific import and medical potential, two public opinion surveys in the past decade clearly show that the great majority of adult Americans do not know the meaning of "genetic engineering" and are ignorant of the ethical and social issues surrounding this activity.[9] And the same appears to be true of adults everywhere. The public is simply not interested, even in a science where one does not need to draw upon mathematics for some reasonable understanding and which is not nearly as conceptually profound as thermodynamics, yet promises to have a far greater impact on the personal lives and well-being of the general public! Here lies the *real* danger of scientific illiteracy; genetic engineering has already spawned groups of modern-day Luddites (neo-Luddites) who prey on the public's ignorance with mostly groundless warnings of dire consequences resulting from attempts to interfere with nature, like producing monsters or even the cloned Alphas, Betas, and Epsilons of Huxley's *Brave New World*.[10] The net result of such movements, whether by charlatans or by self-appointed, but generally ill-informed, guardians of society, is to slow the progress of science and technology, and mark its practitioners as a careless, perhaps even heartless, segment of society.

The storm of protest following Snow's thesis was unprecedented, although it surely had its counterpart in the criticism that raged over Herbert Spencer's suggestion a century earlier that science was the only subject worthy of study.[11] In the main, at least on the issue of communication, Lord Snow was clearly on target: on campus the two groups do not—in most cases, cannot—communicate with one another in their respective subject areas, often not even at a freshman level of sophistication. For the most part they tend to avoid discussing their respective subjects in the academic setting, maintaining a peaceful coexistence that comes apart only when members seek to protect their respective turfs during faculty debates on core curriculum requirements.

Off campus their behavior is somewhat different; in ordinary social interactions between scientists and nonscientists, neither group seems much concerned about the other's lack of understanding of matters outside one's special field. The thing to do when having cocktails with faculty members from other disciplines is simply to listen as experts talk about their current scholarly activities, which often can be very illuminating, or to discuss matters of mutual interest like the failings of deans or central administrations, but to avoid discussing any controversial intellectual subjects because then the playing field becomes uneven. Few scientists can talk in a *scholarly* fashion about Shakespeare and still fewer humanists can talk sensibly about the Second Law. So the participants acknowledge a standoff and most abide by the unwritten rule that one should not challenge a colleague on his or her own turf for fear of being thoroughly embarrassed by the outcome. It is almost as though the respective cultures are saying, "Don't expect me to be literate in science and I won't expect you to be literate in my special field." But their positions do not quite balance, as all faculties easily recognize. Most scientists, if for no other reason than necessity brought on by having to live in a humanistic society, are literate to some degree in the humanities; very few humanists, because they do not regard it as a cultural imperative, are at all literate in science.

In fairness, the question of balance prompts an obvious rejoinder to Snow's challenge, one that appears not to have been raised at the time: why must the two sides be equal? This is not a competition, where to be equitable the contestants ought to be evenly matched. Most academic scientists take the imbalance for granted, perhaps acknowledging the fact that life is not always fair, not even campus life, where other faculty inequities easily outweigh this one. The real issue is one of appearances to impressionable students, who can hardly be expected to embrace science if their nonscience professors are openly indifferent, or even hostile to it.

Thus, while the two-cultures controversy may be dormant it is nevertheless still alive, and promises to remain so despite the efforts of a small group of scientists and humanists who hope to find a common ground. The Society for Literature and Science, formed several years ago, attracts scholars who seek parallels between cultural forces originating in literature that they believe may influence the direction of science (although more likely it is the technology that is culture-driven) and how changes in science are reflected in literary works. However productive such scholarship may seem to those engaged in it, it is difficult to see how showing a remote link between science and literature can bridge the wide gap separating scientists from humanists generally, particularly as most of the researchers in this fledgling field come from the humanities. The walls may begin to crumble only when scholars from the sciences and the humanities are able to penetrate the other domain in a significant fashion, which is certainly not yet the case. It may be recalled that the well-known British literary critic F. R. Leavis, in a lecture delivered three years after Snow's remarks on the two cultures, pronounced Snow's literary abilities as "beneath contempt." Leavis's scientific competence is unknown and matters little, but his intemperate remark is a good example of how heated this controversy became. The Society for Literature and Science reminds one in a way of the current STS (Science/Technology/Society) movement, where the research scholars come mainly from the social sciences and have little direct involvement with science or technology, being concerned chiefly with the impact of these enterprises on society.

▲ PASSIVE RESISTANCE TO SCIENCE

Most science illiterates are merely indifferent, accepting science and technology as essential to modern civilization, more so for their practical end products than for their intellectual appeal. But they do not feel it necessary to be personally involved in science to the point of becoming literate in the subject. We have already pointed out the strange paradox that most science illiterates claim to be interested in science and believe that their basic understanding of science and technology is at least adequate for everyday life (some actually think it quite good), which can only mean that the average individual manages quite well in our society with very little understanding of science.[12] Some, while accepting the benefits of modern technology, deplore its apparent inroads on the tranquillity of life, contrasting the quality of modern existence with memories of the past, evoked either from experience or more often from literature. Most academic humanists, especially the literati, fit this canvas. They prefer not to have their minds "cluttered" with the details of a seemingly unfriendly culture, one far removed from their own—one that seems mechanical and ordered, rather than free-ranging and nonconformist; one that some fear may actually stifle their imagination and restrict their literary freedom. Apparently, it matters not that they see science with a kind of tunnel vision; their perception of it is enough to turn them away. Some, in fact, wear their ignorance of science almost as a badge of honor, saying, in effect, "Don't bother me with the way things are. I prefer to think of them as the way they ought to be."

Passive resistance to science can be as damaging to the science literacy movement as active opposition. After all, who but the faculties of our institutions of higher education are the intellectual role models of society? If most are illiterate in science and do not even try masking their indifference, what, as I asked earlier, must be the effect on students? In a typical liberal arts college, roughly one-quarter of the faculty will be found in mathematics or the natural sciences, with the rest in the humanities and social sciences. The typical nonscience student will spend 15 to 25 percent of his or her time in contact with science and mathematics faculties, the remaining time with nonscience faculties. Hence the attitudes of nonscience faculty members toward science are bound to have some effect on the attitudes of students. It is well known that most students tend to shy away from science and mathematics, and given free choice would generally not elect science courses. The same is true of science majors, most of whom would prefer taking additional courses in their major field than required courses in the humanities and social sciences. Hence were it not for the distribution requirements of most liberal arts colleges, course enrollments would be chaotic and college education would most likely be more one-sided than it is today; science-bound students would avoid the humanities and the nonscience students would avoid both science and mathematics.

▲ ACTIVE RESISTANCE TO SCIENCE

Some individuals admit to being anti-science, viewing the whole enterprise as responsible, either directly or through oversight, for many of the ills of modern civilization, such as pollution, ecological disasters, or military weapons. Here lies a real danger. It is one thing to be neutral or disinterested in science, for which the scientific community may be faulted for having failed to get the proper message across; but it is quite another to be aggressively opposed to science on an agenda that has more to do with the uses of technology than the practice of science, or worse, because it provides a convenient platform for advancing social, political, or private causes. We know that virtually all science-based societal issues are grounded in technology rather than in basic science, that is, in the practical uses to which *society* chooses to apply scientific discoveries. One might choose to challenge the premise that society has any more to say about the technologies that are ultimately developed and become part of our everyday lives than about determining the nature of the research conducted by the scientific community. Yet society does have a measure of control over technology, which it sometimes exercises through referenda on community-based technology issues, or, in the final analysis, can influence through its choice of products in the marketplace. Not so with basic science, unless it happens to be "big science," which requires huge federal funding and hence the support of legislative bodies and their constituents.

Moreover, most societal issues arise from the manufacturing processes used to produce end products rather than from the products themselves, that is, from noxious or hazardous factory effluents, combustion and other waste products contaminating air and water supplies, or thermal loading of rivers and streams used for cooling purposes. This is where serious critics of science might better focus their concerns, rather than condemn all of science in the same breath. It is here that society should have a strong voice, but to try to control the direction of basic science is fruitless and potentially harmful, not only to science but to society itself. Society *must* learn to distinguish between science and technology. Demanding that science avoid doing basic research that *may* in the end result in products or processes harmful to mankind or to the environment is patently absurd; one cannot halt the search for knowledge, particularly where the outcome cannot be predicted. To insist, as do many in the science counterculture movement, that "societal impact" studies be done before undertaking any frontier scientific research, or worse, to espouse the Aristotelian view that science and technology should in *all* respects be controlled by the state, reveals a profound ignorance of the nature of these enterprises. We alluded earlier to basic science as being a "voyage of discovery." If one knows beforehand what the outcome of a given research activity is likely to be, it is hardly *basic* research. It is the unexpected outcome that really propels science forward. Indeed, virtually all *major* milestones in the history of science were unanticipated. Who could have foreseen that the basic studies of Michael Faraday, Hans Christian Oersted, and Heinrich Lenz early in the nineteenth century would eventually give rise to the electric power industry, or that the work of Clerk Maxwell and Heinrich Hertz later in that century would form the basis for all wireless communication, including radio and television? Even more to the point would be the *accidental* discovery of X rays by Roentgen toward the end of the century while investigating an entirely different phenomenon, namely, the effects of cathode rays (i.e., electron beams).

The unexpected is not always beneficial, of course. When in 1905 Einstein put forth his famous mass-energy equivalence ($E = mc^2$) he probably would have been appalled to think that some forty years later it would play such a central role in nuclear explosives, but he had no way of knowing this at the time—and nor did anyone else. In fact, only twenty years later, no less a distinguished physicist than Ernest Rutherford is said to have scoffed at the notion that nuclear power might be obtained from the atom. This inability to forecast the uses to which scientific discoveries may be put strikes at the heart of the problem, pointing up as it does the folly of attempting to control the growth of knowledge and directing society to seek other ways of resolving legitimate public concerns about science and technology.

Most professional critics of science and technology come from outside the scientific community and generally are themselves ignorant of science. Some, in fact, like political theorist Langdon Winner, seem to believe that becoming literate in science brainwashes one into a mold of conformity that lessens the individual's ability to criticize the establishment. Winner, for example, in rejecting the notion that technology critics ought to be well educated in science and technology, states: "Those who suggest a technological education of this kind implicitly ask that one undergo a process of socialization," implying that what he calls an "informed outsider" can be a responsible antagonist.[13] Such rationalization of scientific illiteracy, barely edged in truth, is hardly defensible. Of course it is easier to be critical from the outside looking in, to denounce the unfamiliar out of ignorance, but the danger is that uninformed criticism is likely to be technically wrong and is rarely constructive. Anyone can be a critic—the more ignorant about a subject, probably the more vocal—but if one of the purposes of criticism is to win over the opposition to a particular point of view, naysayers like Winner must come to realize that this can best be accomplished within the system. What society needs are fewer science illiterates who fashion their careers out of criticizing modern technology before receptive audiences in the equally illiterate public, and more skeptics who are literate in science and technology and can therefore target their criticism in a truly constructive manner.

▲ THE SCIENCE COUNTERCULTURE

A third culture, more vocal than the passive science opposition, and apparently growing in presence as well as in numbers, consists of fringe elements that collectively may be termed "science counter-culturists." Some are openly anti-science and anti-technology, others claim not to be anti-science but nevertheless are so by any rational standard, and a third category of people merely wants to remake science in its own image. The origins of the first two groups can be traced almost entirely to real or imagined concerns about the effects of technology on (a) the environment, or (b) the physical, economic, or psychological well-being of society. A particular target of this group appears to be health care and its related medical sciences, where they contend that health care professionals should be paying more attention to medications derived from plants and to alternative modes of medical treatment. The point is well taken. Actually, both the pharmaceutical industry and government agencies have for many years carried on systematic studies of natural plant compounds as potentially useful drugs, with some positive results. Rather than discovering potent disease-specific drugs through this route, however, it is more likely that, starting with the natural substances, clues will be found to improved drug designs in the pharmacology laboratory. As for alternative medicine, which has been plagued in the past by subjective or unsubstantiated claims of miraculous cures or remissions, this too has now achieved some degree of legitimacy, at least as measured by the fact that a new federal agency has been set up to formally investigate such "New Age" alternatives.

The third group, however, is motivated only partly by a desire for social reconstruction; most of all, in the guise of a human-centered natural philosophy, it seeks to overturn the very foundation of science, namely, respect for truth and rational thought. In fact, these counter-culturists, or "postmodernists," as some prefer to be known, reject rationalism as the best way of dealing with the science-society interface, or even of trying to understand nature. Science and technology have always had their share of critics, some serious-minded, others opportunistic, but these were nothing compared to those now swarming about science (particularly around the new Science/Technology/Society movement to be discussed later) like honeybees about a newfound source of nectar. The reason seems obvious: many now look upon science, not as a force for progress, but as the source of many of the ills of modern civilization—a convenient vehicle for criticism and confrontation.

They are overwhelmed by the complexity of modern science, reject it and adopt in its place various self-centered philosophies that appear to provide them with a gratifying "spiritual" awareness of the universe. It is little more than an escape from reality. Under the collective umbrella of the "New Age" movement, these doctrines range from the more profound but fruitless efforts of Fritjof Capra to link Eastern mysticism to modern physics[14] or of Marilyn Ferguson to create a new social consciousness of the human potential[15] through the paranormal delusions of crystal energy, mysticism, the "psychic powers" of Uri Geller, and creative visualization, to name only a few, to the bizarre channeling claims of the scientifically inept but otherwise highly accomplished actress Shirley MacLaine. Both Capra and Ferguson are thoughtful exponents of seemingly plausible but highly speculative concepts. Capra, a theoretical physicist who also became thoroughly immersed in Eastern mysticism, particularly the religious philosophy of Taoism, believed he saw parallels between some of the philosophical paradoxes of modern physics, particularly in the domains of quantum theory and relativity theory, and the way that Eastern mystics view the world. A major problem with Capra's thesis is that in order to draw such parallels one must abandon the traditional notions of logic and reason, inherent in Western philosophy and expressed in common language, in favor of the ethereal explanations of reality, transcending ordinary language, that characterize the Eastern philosophy embraced by Capra. Seeking to account for what seems mysterious in science through theosophy and symbolism, though it may dramatize the limitations of ordinary language, does little for science in terms of providing better models of nature, or for students seeking to understand these models.

On the other hand, Ferguson does not propose a specific mystical view of nature, but rather presents an elaborate vision of our future consciousness brought about by anticipated advances in brain research, and presumably in cognitive science. Ferguson asserts that this expansion of the human potential is already underway through an informal underground network (the "Aquarian conspiracy") seeking to create a new kind of society better prepared to meet the intellectual challenges of the future—the knowledge revolution that many feel is already upon us. She provides an optimistic view of a future in which people take charge of their own destiny through altered states of consciousness, whether derived from newly developed styles of education, spiritual encounters, meditation, or even from experiences with mysticism and the occult. That great changes in our understanding of the brain will emerge from the laboratory is indisputable, but whether Ferguson's utopian outlook on how such new knowledge may influence the structure and workings of our society can be considered accurate is highly questionable.

As for the others—the channelers, the spoon benders, the illusionists, the psychics, the astrologers, the readers of crystal balls, tea leaves, and tarot cards—from the scientific point of view they are either intentional entertainers, self-deluding individuals, or simply outright frauds and charlatans.[16] The last of these obviously comprise still another obstacle to improving the scientific literacy of society at large.

▲ THE ANTI-SCIENCE/TECHNOLOGY MOVEMENT

There is a new wave of anti-science/technology, one that challenges the basic fabric of science and seeks to capture the minds of the scientifically illiterate. The fact that many of those who seek to counter the inroads of science and technology on modern life are basically illiterate in these enterprises is small consolation, for so too is most of the society they seek to persuade.

One group of anti-technologists, whose members call themselves "neo-Luddites," began with the premise that all technologies are political and intrinsically harmful to mind or body, and actually favors dismantling the nuclear, chemical, genetic engineering, television, electromagnetic, and computer technologies, among others.[17] They would replace these technologies with newly created ones that, in their words, "are of a scale and structure that make them understandable to the people who use them and are affected by them." Presumably, if society went back to the hand-tool days of more than two centuries ago, prior to the industrial revolution, some neo-Luddites might be happier, but few others would. Think how degrading this rationale must appear to the society they seek to "protect." They are saying, in effect, that since the general public does not understand modern technology, rather than seek to raise its level of understanding we should lower the level of technology.

The term "Luddite" stems from an early nineteenth-century labor movement in England, when the industrial revolution was in full swing and skilled hand labor was rapidly being replaced by machines. In protest, one Ned Ludd led a number of his fellow workers on a reckless, machine-smashing rampage. Today's Luddites do not resort to such violent behavior, of course; the more moderate call instead for a critical study of new technologies before they become fully developed, based on the arguable premise that unless proven otherwise, *all* technologies should be considered inherently dangerous to man or the environment. Many, however, while professing not to be anti-technology, nevertheless readily justify the original Luddism, with its destruction of factories and labor-saving machinery, as responsible collective activism. Moreover, posing a clear threat to science education, they reject rationality as the cornerstone, not only of responsible behavior, but also of science and technology. Shrinking the dimensions of a discipline within borders "understandable" to everyone is among the surest ways to stifle creativity and the growth of knowledge.

Should the scientific literacy movement, and particularly its STS component, which appears to be a special target of such groups and individuals, ever become dominated by them, the movement would be severely handicapped, for the science and technology communities would then lose whatever interest they presently have in helping to develop such literacy. Nevertheless, it would be folly to assume that simply because science has made such remarkable advances since the Renaissance, and through technology now influences the lives of all members of industrial societies, it is *uniformly* acknowledged as a great boon to the world. Many would argue this point, not merely on the ground that unrestrained technology has often proved harmful to the environment, or even to certain segments of society, but also because some find it a spiritless influence on what is basically a humanistic (nonscientific) culture. Prominent among such science counterculturists are a number of scholars, mostly sociologists and political scientists who either seek a better understanding of the social mechanics of science (and of scientists), or, having already convinced themselves that science, as conventionally practiced, does not meet the intellectual needs of most of society, actually propose other ways of practicing science or of regarding natural phenomena.

Both the anti-technologist Jacques Ellul[18] and the social critic Lewis Mumford,[19] whom we shall meet again later on the question of expert advice, have warned of the dehumanizing potential of technology. Ellul goes further to deplore the danger of society being dominated by rationality and logic to the exclusion of spontaneity and creativity. Ellul's main argument is that the "technique" underlying technological advances—that is, the process by which new technologies and machines are brought into use—has become so routine and stereotyped as to depersonalize the workplace and even the school. The ultimate danger, he foresees, is that all human activity will fall into such a mechanical mode and everyone becomes an automaton. The American critic Theodore Roszak, clearly influenced by Ellul, spoke compellingly for the counterculture as he appealed to the youth of the nation at a time of its general disaffection with the social establishment.[20] While addressing a more specific audience, his thesis was much the same: science and technology were depriving society more and more of human values and freedom of thought. Even though one may strongly disagree with any belief that rejects rationality as the fundamental basis for civilized individual behavior, and particularly for the practice of science, it must be appreciated that both Ellul and Roszak reflected the thinking of a large segment of society, which we know to be guided more by emotion than by reason, and for whom superstition, pseudo-science, and the occult often strike a more resounding chord than does scientific reasoning. One also derives from their writings an uneasy feeling that they do not see clearly the distinction between science and technology, lumping these activities together as one and the same, as do most people outside the science/technology community.

On the other hand, social philosopher Paul Feyeraband, a one-time student of science, took a far more negative view of the science/technology enterprise.[21] His was not a reflection of contemporary societal attitudes as much as a personal conviction that the practice of science, and even some of its foundations, is seriously flawed. He distrusts scientists and educators—in fact all intellectuals, particularly liberal intellectuals—as well as rationalism. He is fiery in his defense of astrology and vehement in attacking its critics; and he believes that science, which he characterizes as a threat to democracy, must be supervised by the general population, however poorly they may do so. How can one react to such extreme views by a scholar? To a practicing scientist, any debate that starts off by rejecting rationalism is already highly suspect and to be avoided, for then there is no logical method of refuting the argument, at least not one acceptable to an irrationalist. In this case, however, Feyerabend manages to demolish his own position by presenting what he regards as *scientific evidence* in support of astrology. The evidence he presents is so weak, so speculative, and so filled with hand-waving as to question his understanding of the meaning of scientific evidence, of what science is, and by extension his judgment generally on the role of science in a democracy. Yet unfortunately, although perhaps understandably, his influence on the science counterculture community appears to be substantial.

Political theorist Langdon Winner[22] and philosopher Joseph Turner[23] exemplify a different aspect of the science counterculture. Their concern is more with the social responsibilities of science and technology, with fitting these into a democratic society more than with characterizing the nature of these enterprises or speculating on their ultimate effect on human affairs. Winner, for example, does not agree with Ellul and Roszak that technology is running amok and unless brought under control will eventually acquire a momentum and determination of its own that will somehow enslave mankind. Winner refers to such a runaway process as "autonomous technology." Both Winner and Turner speak of *democratizing* science, particularly technology. These are not radical critics of technology but rather they raise serious questions about whether society is prepared to deal intelligently with the technology explosion that has been thrust upon it, or are new institutions and modes of thought needed to bring the two into harmony. Obviously this is an important element of the science/society dialogue, but a basic problem underlying all such efforts to accommodate science and technology to contemporary society is the difficulty of dealing with the issue of *democratizing* these enterprises in the abstract. Instead, because different segments of society may have conflicting views of how technology is best adapted to societal needs, ranging from absolute control of technology by the community to the more limited controls imposed by existing legislation in the industrialized nations, one must challenge the premise by asking in whose image science and technology should be democratized. For example, most neo-Luddites and anti-scientists think of democratizing technology according to their personal desires, not necessarily for the overall good of society. They regard the democratization of science as meaning not only that society should control the *direction* of science, if that were possible, but also that all personal theories and private views of the natural world are equally valid, ignoring the process by which we have learned what we know about science.

This last point, namely, constructing a personal view of the world, is one of the basic tenets of the "constructivist" or "postmodernist" school of thought, comprised largely of social scientists who assert that contrary to the scientists' (mainly positivist) view, nature cannot be studied objectively. Instead, every observer is biased in some fashion that limits his or her ability to be wholly objective. Typical of this group of counterculturists are sociologists Steve Woolgar and Bruno Latour, both of whom seem prepared to ignore what success the established methods of science have produced and to discard these in favor of unproved assumptions. Woolgar seeks to discredit science as a means of obtaining reliable knowledge about natural phenomena. He calls upon a theoretical construct popular among constructivists, known as the Sociology of Scientific Knowledge (SSK), to support his view that understanding nature (in the scientific sense) is not as important as devising alternate explanations for it that are more pleasing to the social scientists.[24] And Latour carries the relativist-constructivist position to the sublime by asserting that science consists mainly in the *power* of its proponents—that it matters little how shaky one's proof may be for a given statement about nature as long as others can be made to believe it—by any possible means.[25] This is characteristic of the "scientific theories" advanced by the constructivists. Unfortunately, many nonscience students are receptive to such thinking because no norms or achievement levels are imposed. More will be said later on the counterculture movement.

▲ ANTI-SCIENCE FRINGE GROUPS

There is little doubt that many industries (and governments), knowingly or otherwise, have acted recklessly with the environment; and there is no question that the products of science have often been used to the detriment of society rather than its benefit. We know that science and technology are often forced to dance, however reluctantly, to the tune of government and industry funding. Hence there are good reasons for mistrust and genuine criticism. The ecologists and environmentalists, and those groups genuinely concerned with the health and safety of humans and animals, to the extent that they appeal to reason rather than emotion, serve society well by calling attention to obvious abuses. But to generalize from such offenses, to use these arguments as the bases for blind condemnation of science and technology, or as the rationale for exploiting some covert agendas of anti-science and counterculture groups, would in the end, like throwing out the baby with the bath water, be the greater evil. The fact that most of those who deplore the inroads of science and technology on modern life are basically illiterate in these fields is of little consequence, for so too, we know, is 90 to 95 percent of society. Hence the arguments used by such groups, unreasoned though these may seem to one trained in science, nevertheless often carry some small grain of truth, enough at least to gain the attention of uncritical audiences.

All this might be dismissed as pure nonsense were it not for the fact that to win converts the neo-Luddites must capitalize on the very same public ignorance of technology that prompts it to propose such absurd remedies. The appeal of the neo-Luddite movement, as of all the modern anti-science or anti-technology fringe movements, is to fear and emotion, not to the reasoned dialogue and debate that supposedly forms the cornerstone of science. Speculation, hearsay, anecdotal "evidence," half-truths, wishful thinking—these are the stock in trade of the neo-Luddites and their supporters. They *capitalize* on doubt rather than seek to resolve it. Preying on fears of the unknown and ignoring risk-benefit assessments are standard practice of these modern-day Luddites, many of whom seem to view technology as a convenient vehicle for attacking government, big business, and science itself. For example, among the technologies that psychologist C. Glendinning claims the neo-Luddite movement favors dismantling are genetic engineering technologies, "which create dangerous mutagens that when released into the biosphere threaten us with unprecedented risks,"[26] And according to the same manifesto, also to be dismantled are electromagnetic technologies, "whose radiation alters the natural electrical dynamic of living beings, causing stress and disease." No evidence is presented that these are indeed real risks, or that the precautionary measures required by law are inadequate to prevent such occurrences. No weight is given to all the studies designed to determine the effects of electromagnetic radiation on higher organisms (above the cellular level), which have thus far proved negative. Nor are the benefits of these technologies cited, which might permit one to judge which outweighs the other. Instead, simply raising the specter of monsters or new diseases, or simply fear of the unknown, invariably finds some receptive audience among the general public.

"The end justifies the means" appears to be the guiding principle of all such movements. Yet against all such fuzzy reasoning and its affinity to uncritical thinking, scientists and science educators can appeal only to objectivity and respect for verifiable truths. Unfortunately, as we all know, when reason conflicts with emotion, or even with what many would term "common sense," reason frequently loses out, making it easier for these fringe elements to strike responsive chords in the minds of many in contemporary society.

Can all this be dismissed as a harmless search for self-image, or does the anti-science movement constitute a real danger to society? Probably some of both. Unfortunately, it cannot be dismissed out of hand, for it is not merely a contemporary aberration but a recurring phenomenon. As Gerald Holton points out in his illuminating work on anti-science, "history has shown repeatedly that a dissatisfaction with science and its view of the world can turn into a rage that links up with far more sinister movements."[27]

Postmodernists and "science-bashers" generally are taken sharply to task by Gross and Levitt in their critical analysis of what they term "the academic left."[28] These are the scholars, mainly humanists and social scientists, who, in casting about for new fields of scholarship to measure against their own concepts of knowledge, have focused on science. Because their ignorance of science makes it impossible for them to contribute effectively to the field itself, they tend to resort instead to deconstruction, unscholarly criticism, and the social restructuring of science. While Gross and Levitt recognize such attempts to *lecture* the scientific community as amateurish and mostly ineffectual, they nevertheless believe it important to unmask these critics of science for what they appear to be: a group of otherwise diligent scholars clumsily playing with science almost like an infant with a newfound toy.

▲ THE ANIMAL RIGHTS MOVEMENT

There are others that could be numbered among the active counterculture groups focusing their attacks on science and technology, some because they are genuinely concerned over the adverse effects of specific technologies or for humane considerations, but others mainly for political or commercial reasons. Among the former we find the highly visible animal *rights* (as distinct from *welfare*) group, PETA (People for the Ethical Treatment of Animals), which, while effective in many of its efforts to improve the treatment of animals in scientific research, resorts to the same distortions and generalizations that are characteristic of the neo-Luddites—and more; violence and terrorism are considered by the more activist members of this organization to be justifiable means of achieving their goals. While never directly implicated in illegal activities, its more militant members applaud when they are conducted by a kindred underground activist group known as the Animal Liberation Front (ALF). Animal welfare groups, humane societies, and anti-vivisection organizations have been around for more than a century, working quietly and generally within the law to bring about the changes they felt were needed. They were reasonably successful in improving conditions for working animals, show animals, and household pets, although many would argue whether anti-vivisection groups served a useful secular purpose. But they had limited success when it came to laboratory animals, and PETA's more aggressive posture, in effect putting animal rights above perceived human needs, and condemned by most scientists for its tactics, has proved far more effective.

The animal *welfare* movement has always acknowledged the need to use animals in biomedical research, insisting only that such use be responsible and humane. Animal *rights* groups, on the other hand, argue on moral grounds that all such use should be abolished. Since the 1950s concerns over the humane treatment, if not the actual use of animals in any way, have grown, and the scientific community now concedes that it is a force to be reckoned with. Here again, however, it is not a question of debating the issue on scientific grounds, of convincing PETA that animal research is necessary to good biomedical science, for PETA's position at the outset is that (a) as used in research animals cannot provide useful information, and (b) in any event, alternative methods can be found to provide the same information. Biomedical researchers, while conceding that for some purposes cell cultures may yield information equivalent to that obtained from animals, insist that there is a point at which there is no viable substitute except to experiment on humans. Never mind that the number of dogs and cats having to be destroyed annually by humane societies is far greater than the number used in research, PETA (and the anti-vivisectionists) cling to their illogical conviction that animals are not needed in biomedical research and should not be used (dead or alive) merely to gain new knowledge.

Obviously, PETA does not seek to persuade the scientific community of its position, but works around it to target the community at large, using unproved claims, effective public relations and media "events," and even outright deception and exaggeration to bolster their position. PETA takes its case to the "caring" members of the public and through them exerts pressure on industry. The cosmetics industry has already pretty much caved in to PETA's strategy of commercial intimidation, with many companies agreeing to stop animal testing and some seeking a substitute for the "Draize eye irritancy test" used on rabbits. Here, one could argue that using animals is really not an essential need in developing and testing cosmetics and might easily give way to other, more benign methods of determining product safety. However, the same cannot be said of most biomedical research or teaching, on which PETA's tactics have already had a chilling effect because of the wanton destruction of some laboratories and animal facilities by animal rights supporters. Bitterness has developed at both ends of the spectrum. Somehow, a compromise position will have to be found, one that is likely to rely upon strict government regulations on the use of animals in research. Predictably, neither side will be fully satisfied with the solution.

A primary target of the animal rights groups is the classroom, where they can generally find a receptive audience of teenagers who are particularly sensitive to the welfare of animals. But the science and science education communities have begun to fight back, through special committees formed by various bioscience and medical associations, educational literature, legislatures, and the courts, seeking to bring some balance to the issue. It promises to be many years, how-ever, before most people fully recognize that some societal needs tran-scend our natural concerns for animals. So destructive of animal research facilities has been the ALF and related activist groups in recent years that Congress enacted a bill in 1992 (the Animal Enterprise Pro-tection Act) providing severe penalties for such criminal activities.

The growth of the animal rights movement is a classic example of our failure to convince students (and the public generally) of the impor-tance of reasoning with the mind instead of *solely* with the heart. What-ever efforts we make in our schools and colleges to develop habits of the mind, it is obvious that in the real world emotion is often a stronger influence than reason, that rational "scientific" arguments frequently do not carry the day. The answer cannot simply be further science requirements or better science teaching, factors that are so often cited as the keys to scientific literacy. The dilemma actually has little to do with science except that since Dewey's time it became fashionable to use this discipline as the model of critical thinking. But adopting sci-ence as the logical medium for instilling in students Dewey's much admired "scientific habits of the mind" has not been successful. It may be that other disciplines, such as philosophy or law, would be more effective in this respect if they introduced appropriate courses at the high school level.

▲   DELUDING THE PUBLIC

Perhaps the best example of an organization that has no special agenda but seems to land on any science or technology movement that can be made an issue of is Jeremy Rifkin's Foundation on Economic Trends. Rifkin's philosophy appears to be, at the very least, to maintain the technological status quo in the United States, and to use the courts to help achieve that end. He is a self-appointed "professional" guardian of people's rights who knows the optimal issues to choose for maximum public relations effect, and knows too that the best way to get the pub-licity he needs to keep him in the public eye is to litigate. Rifkin's ra-tionale, like that of the neo-Luddites generally, is that society is at the mercy of technology and hence any new technology should be looked upon with suspicion and cast aside if there is even the smallest risk attached to it. The problem is that the risks envisioned by Rifkin are generally based on extrapolations from remotely related experiences, if not on pure fantasy and speculation, rarely on objective calculations, even in those cases where enough data are available on which to base reliable risk/benefit calculations.

With its limited knowledge of science and of how the scientific enterprise works, the Rifkin camp easily adopts false premises to support certain preconceived conclusions, knowing that its target audiences also lack training in science or logic. For example, in his 1990 suit for an injunction against NASA's proposed launch of the *Galileo* spacecraft because it was to carry a plutonium oxide energy source, Rifkin and his co-plaintiffs contended that the chance of an accidental plutonium release over Florida was one in 430, a number that appeared to have no basis in fact and differed grossly from the conclusions of risk analyses performed by NASA and others. A special target of Rifkin's foundation is the biotechnology industry, which sometimes uses bioengineered living organisms, thereby leaving it open to often wild speculation about possible harmful effects. They make good use of the classical debating technique of confusing the *possible* with the *probable,* and they take full advantage of the uncertainties in scientific knowledge. That is, we all know that the best a scientist can do is assign a probability, not a certainty, to something happening. Thus, even though the odds *against* a harmful effect occurring may be huge in a given instance, there still remains some chance of its happening. However small the estimated risk, even if smaller than the risk of crossing a street, for example, or being struck by lightning, this provides Rifkin an entering wedge to argue against introducing new technologies on the grounds that, since the scientific community cannot *guarantee* the outcome of its activities, scientists should not be trusted to deal with the moral or ethical problems that may result from their work. His conclusion, namely, not trusting scientists to decide the moral and ethical problems arising from their endeavors, may be valid, but certainly not for the reason stated.

As a last resort, Rifkin is always able to fall back on the rationale that his organization is working for the "public good," or at least their version of such good, and therefore the end justifies the means. The end they claim to seek is the control of science and technology by society, the same as the goal of the neo-Luddites and other social reformers. The question, of course, is what shall be meant by "control" of these enterprises. If meaningful (and viable) reforms to attain their goal were put forth by Rifkin and others, rather than their extremist propaganda and broad-brush charges of elitism in the scientific community, a fruitful dialogue might result. Otherwise, there is little hope that an accommodation can be reached between the two.

There is one important difference between Rifkin and most professed neo-Luddites: where the latter would demolish existing technologies to suit their particular worldview, Rifkin seeks only to prevent or at least to delay new technologies (or new applications of known technologies) from coming on line, particularly in the field of biotechnology. He is saying, in effect, that technology has already done sufficient damage to the planet, which may be true enough, and therefore it is time for society to put a stop to the rapid growth of technology or at least slow it down to the point where the public can effectively deal with it. But their "scare tactics" are much the same: the more they can alarm the public, the more support they are able to gain from it.

And the formula does seem to work. It works because so many individuals in a democratic society actually relish confrontational issues, especially when the other party to the encounter is big business, government, or specialist groups like lawyers, doctors, or scientists and engineers. College students seem to be particularly attuned to such issues, often as a reaction to authority, and Rifkin takes full advantage of this through frequent appearances on college campuses. Using the courts to help achieve his ends also works to his advantage, at least up to a point. Rifkin's lawsuits have managed to delay for years a number of experiments in biotechnology, sometimes forcing small companies to back down for fear of costly litigation (as on the issue of somatotropin in the milk of cows treated with this apparently harmless hormone). Were it not for the fact that he is able, through such legal maneuvers, to publicize his social and philosophical concerns, he would be just another voice in the crowded arena of professional activists and deliberate obstructionists. But the courts can be his ally only for a while; eventually, the lawsuits are resolved, more often on the side of rationality than emotion, leaving behind only legal costs, frustrated scientists, and, in the final analysis, possibly some cost in human life or suffering resulting from delays in conducting experimental trials of new developments in biotechnology.

Despite the many problems they create, not only in the eyes of industry but also for science education, which must deal with their often irrational, unscientific methods, the fact is that the world needs its "Rifkins"—watchdogs who can keep the scientific establishment on its toes; not Rifkin clones, to be sure, but *responsible* and hopefully impartial individuals who are as knowledgeable in the science and technology they find fault with as in the political arena. Unfortunately, responsible whistle-blowers are very scarce. A method must be found to encourage the entry of such individuals into the technological/political process, for bringing a few organized groups of *respected* technology critics into the debate would help both to ease the concerns of society and diminish whatever influence organizations like the Foundation on Economic Trends may have on an unsuspecting public. We shall have more to say later on the subject of using experts on scientific matters.

▲ THE "POWER" OF THE PRESS

One might assume that the popular press, in its role as chief public informant via print in the United States, would be a strong advocate of efforts to alert readers to the dangers of pseudoscientific claims, but regrettably this is not the case. The little coverage of science found in the popular news media is more likely to be sensational than educational, or equated with astrology or other forms of the occult, and those that are informative seem to attract readers mainly when they deal with personal health. There appears to be a strong compulsion on the part of science writers to overstate science news, to emphasize the human interest or other nonscientific aspects of a science story rather than try to explain its scientific significance. While perhaps understandable from the reportorial point of view, it more often than not portrays an inaccurate image of science and its practitioners. Most newspaper editors and writers believe that the function of the press is to cover what is "newsworthy," not necessarily educational. A few enlightened newspapers, to their credit, try to do both, but compared to what could be done by the popular press in the area of science and technology, this amounts to little more than a drop in the bucket.

Unfortunately, science education (or any kind of education) is not what most of the public want from their daily newspapers, as publishers are well aware, but at the very least readers should expect accuracy and freedom from sensationalism and bias in what little science coverage can be found in the press. Here, newspaper editors obviously play a major role. Perhaps more disturbing than the results of surveys of scientific literacy in the general public are studies of the knowledge and attitudes of newspaper editors on science.

In 1987, for example, shortly after the most recent creation-evolution controversy was finally resolved by the U.S. Supreme Court, a questionnaire was distributed by Michael Zimmerman to the senior news executive at each of the then 1,563 daily newspapers in the United States.[29] The first part of the questionnaire consisted of a number of declarative statements requiring the respondents to indicate the degree to which they agreed or disagreed with each one. A total of 534 responses were received (34 percent of the newspapers surveyed), which is a remarkable return. Bear in mind that the purpose of this survey was to examine the attitudes of newspaper editors on creation science, and in this connection one should realize that repeated surveys of the general public had shown that the vast majority of people are at least sympathetic to "creation science." Previously, according to some public surveys, as many as 80 percent of those questioned wanted creationism brought into the public school classroom. Even worse, studies of college and university students have shown that a large majority favor the introduction of "creation science" in the public schools, and a survey of high school biology teachers in Ohio at about the time that the creation controversy was at its height found that at least 15 percent of the biology courses offered in that state present "creation science" in a favorable light. Against this backdrop it is perhaps not surprising to find a large number of newspaper editors unwilling to disagree *strongly* with some of the clearly erroneous tenets of "creation science." For example, only half the editors surveyed disagreed strongly with the statement "Dinosaurs and humans lived contemporaneously." Similarly, only 57 percent disagreed strongly with the statement that "Every word in the Bible is true," and only 41 percent disagreed strongly with the statement that "Adam and Eve were actual people," roughly the same proportions as found in the general public. Also to the point was the finding that of those editors who had devoted any space at all to the controversy (about 60 percent of the total), fully three-quarters had given equal or more space to creation science than to evolution.

These findings tell a disturbing story. Two of the basic dogmas of creation science are the coexistence of humans and dinosaurs and a young Earth. To find the views of half our newspaper editors at odds with those of the scientific community—to find them thinking with their emotions instead of their minds—is distressing enough, but to find them also just as ignorant of some of the most basic facts of nature as their readers is inexcusable, especially as they have far more control over the *continuing* science education of our adult population than does the academic community.

That the power of the press is a factor in the scientific illiteracy of our adult society and that it has the capacity for reversing the current trend are surely worth considering. If awareness of science by adults is to be a realistic goal, the only possible route is through informal education, particularly by the mass media, which have an opportunity here to effect some change in the way that society looks upon science. It also holds the key to reinforcement of science facts and concepts for educated adults in the future, and to the exposure of scientific frauds. Unfortunately, as we have seen, the popular press for the most part devotes space only to the "newsworthy" in science, the spectacular, the amazing, the commercial aspects of science, not to the task of helping to educate the public. Obviously, if a choice had to be made, any newspaper editor would throw out a science article in favor of some fast-breaking news story, but would the editor consider sacrificing the comic page or the daily horoscope? It all boils down to the question of where a society places its priorities, for in the end one must assume that in terms of style and coverage newspapers ultimately respond to their readers' wishes.

If all newspapers gave as much space to science columns on a regular basis as they devote to the pseudoscience of astrology, it is conceivable that our society would not be as illiterate in science as it is today. Of the roughly 1,600 daily newspapers published in the United States, some 1,400 carry regular (daily) astrology columns, and of these only a handful (fewer than fifty) print disclaimers along with their astrology columns to the effect that horoscopes are not based upon scientific fact, are presented only for entertainment and not as guides to the conduct of one's daily life.[30] Freedom of the press is a cherished right in a democracy, but with freedom come responsibilities, among these being the obligation to use the enormous power of the mass media for the overall good of society. Promoting pseudoscience in the guise of astrology, innocent though it may seem, cannot be counted as contributing to such good.

▲ CHAPTER 6

Recent Approaches to "Scientific Literacy"

What's past is prologue.
—Shakespeare (*The Tempest*, Act II)

▲   A number of new precollege curriculum development programs have been launched in the past few years, mostly with National Science Foundation support, in a renewed effort to capture the elusive scientific literacy. All are on a local or regional level, presumably to be replicated nationwide if successful, but a few are already sufficiently large in terms of financial and personnel commitments to qualify as national ventures. To place these curriculum reforms into proper perspective, it may be helpful to look first at some earlier attempts to restructure the science curricula at the elementary and secondary school levels.

▲ ELEMENTARY SCHOOL SCIENCE

A widely held truism in science education is that most elementary school children are captivated by *hands-on* science activities. Science is basically culture-free, at least at the elementary school level; hence there are few disadvantaged children when it comes to doing "science" in the elementary grades. Since the 1950s there had been growing evidence that much more could be accomplished at this level than was believed possible in the past. The motivation and ability of children in the primary grades to deal with scientific concepts appeared to have been grossly underestimated, and many scientists and educators became convinced that the elementary school was the arena in which the greatest impact could be made in science education. This view was strengthened by the belief that many young children will have developed their thinking patterns by the age of twelve. Hence, in these formative years, when the

129

natural curiosity of children about the world around them is at its peak, and minds are so receptive to new ideas, it was thought possible to develop a foundation in science that would remain a permanent part of the individual's intellectual life, one that would serve students first during their school years and perhaps even later as responsible adult members of society. If little else, it should be possible, many believed, to avoid the misconceptions in science that seem to plague so many students (and adults). The elementary school offered a challenging opportunity for still another reason: because it had fewer vested interests and constituencies, it was considered more receptive to major curricular experimentation than the secondary school.

At least this was the basic premise during the 1960s and 1970s, when the first major round of elementary curriculum reform took place. Such nationally funded programs as ESS, SAPA, SCIS, and COPES (the "alphabet soup" programs),[1] hailed by science educators then as the modern wave of science education, quickly gained acceptance in a large number of elementary schools across the country. Textbooks, teachers' guides, equipment kits, in-service programs, all designed to implement the educational process, became the norm as the science education community sought for a more meaningful introduction to science in the elementary schools. A common element in all these programs was the involvement of children (and teachers) in hands-on activities. In this regard the programs represented a great departure from most of the traditional elementary science *reading* programs then in vogue.

Another departure common to these programs was an emphasis on science process and on the structure of science. In a sense, the programs sought to present the scientific enterprise as seen more nearly through the eyes of the scientist. Indeed, a characteristic feature of most curriculum reform projects of that period was a heavy involvement of academic scientists in their design; some science educators, in fact, claimed that the programs were overly influenced by university scientists and failed because of that. On the other hand, an often-voiced complaint of the scientists was that the learning theories and educational philosophies underlying some of the projects led to their ultimate demise. Probably there is some truth in both, but the more likely reasons, we shall see, had little to do with curriculum content.

A word is in order on the use of "hands-on" as a descriptor of certain science programs, particularly as some critics of science education seem fond of suggesting (perhaps facetiously) that hands-on may be taken to mean "minds-off." Actually, its meaning should be obvious:

it came into general use with the alphabet soup programs at a time when elementary school science consisted mainly of reading *about* science, with occasional teacher demonstrations and observations of some classroom plants and animals. The new programs had children handling simple equipment, doing experiments, making measurements, recording data—in short, experiencing science directly rather than reading about it, or worse, listening to teachers read about it. These were intended to be directed activities, not mindless playing with objects, and sought to involve all the children's faculties: hands, eyes, ears, nose, *and* minds. Hands-on means activity centered, with the children personally involved in the activities. To suggest otherwise is little more than professional posturing. Whether these hands-on programs were fully effective is another matter to be discussed further, but any shortcomings they may have had cannot be attributed to their hands-on character. In fact, "hands-on" is now the catchword of virtually all elementary curriculum developments, along with "inquiry-based" (previously known as "discovery-based") science.

It was also in the '50s and '60s, we have seen, that the notion of "scientific literacy" as a goal of science education began to surface. The term appears to have first been used by both Paul Hurd and Richard McCurdy in 1958.[2] Previously, the major goals of science education had a more practical bent, in terms of both curriculum and desired outcomes. The immediate post–World War II period (the late '40s and early '50s) saw a rapid growth in American peacetime industry, with a corresponding need for many more scientists, engineers, and allied professionals, and obviously more science teachers. Thus, the emphasis then was on getting more students into science-related careers. As for those students not interested in scientific careers, meaning the vast majority, some exposure to science had long been thought necessary, both in high school and college, on the theory that a discipline so prominent in human affairs deserved to be part of the general education of all students. Whatever the intended meaning of scientific literacy, which at the time certainly was not clearly defined in an operational sense (nor is it now), it was believed that the new elementary science programs might lead students toward this elusive goal more effectively than the traditional science (reading) programs then in use. This failed to occur, as we now know. Nevertheless, curriculum reform efforts continue in the hope that a magic formula will somehow be found to painlessly transform the United States into a nation of scientific literates.

In part, the curriculum reforms of the 1960s and 1970s were a reaction to the practical science of the previous decades. Beginning in the high schools and eventually moving down to the elementary grades, developers sought to bring more of the theoretical, conceptual structure of science into the classroom, reflecting the view of the scientific community that the facts of science contribute far less toward understanding the nature of the scientific enterprise than does the process of accounting for these facts. As a result, where prior science programs emphasized content (facts, laws, formulas, problem-solving, etc.), the new programs stressed the ideas and process of science (conceptual schemes, theories, scientific [analytical] reasoning, logic, creativity, etc.). In fact, one of the major criticisms leveled at the new programs during the 1960s (e.g., by Ausubel,[3] Fischler,[4] and Atkin[5]) was that they put too much stress on science process at the expense of content. But after all, wasn't this *real* science as seen by the scientist? How else could one meaningfully portray the essence of this enterprise, particularly at a time when the aftermath of the war, followed by the shock of Sputnik, cast so strong a spotlight on science education? It was a belief that the new programs more accurately represented the true nature of science, coupled with an increasing awareness of the impact of science on society, that fueled the hope that widespread scientific literacy might be an attainable goal. This hope was never realized, but for reasons that cannot be linked directly to faults of the curriculum.

The millions of students who were exposed to the new hands-on programs of the 1960s have now entered the mainstream of society, prompting the obvious question of whether they are more knowledgeable about science matters than were students of the 1940s and 1950s, whose exposure to science in the elementary grades was mainly through reading programs. In other words, did the carefully designed "alphabet soup" programs, with their presumed solid grounding in learning theory, science process, and the foundations of science, have a discernible lasting impact on the students (now adults) who were exposed to them? However more gratifying these new programs may have been to their developers, to the students who were exposed to them at the time, and in many cases to their teachers, there is no convincing evidence that our young adults today are any more knowledgeable or sophisticated in science than were previous generations.

Unfortunately, no comprehensive longitudinal or retrospective studies are available to quantify this conclusion; hence one must resort to overall impressions from the few available surveys of adult literacy in science, which appears to be no better today than it was a generation or more ago. But there is more telling evidence than this: had the early reforms been even moderately successful in producing a scientifically literate public, we would not have witnessed the clamor of the '80s for still another round of reforms in science education. For example, among the many commission reports of the early 1980s, the National Science Board Commission urged the development of "new science curricula that incorporate appropriate scientific and technical knowledge and are oriented toward practical issues."[6]

Perhaps retention of science process and concepts into adulthood is not a fair measure of the value of an elementary science curriculum, although in the final analysis the mark of a literate individual should be how well he or she fits into, and contributes to the good of, contemporary society, not merely how well the individual performed in school. At any rate, it may be instructive to look briefly at the results of the standard methods of evaluating innovative curricula, as these were applied to the new elementary science curricula of the '60s and '70s— that is, by comparing the performance of experimental groups with that of controls on a contemporaneous basis, meaning soon after exposure to one of the new curricula. A number of postmortem studies were done in the '80s, with similar results: if anything, the overall positive effects of the new curricula were marginal; they had no *profound* effect on student performance.[7] Whatever positive results were observed, while considered statistically significant, were not very impressive, and there is always a danger in looking to statistical significance alone, in the absence of other compelling evidence, as the measure of worth of an educational experiment.

Obviously, if a new curriculum exhibited a "spectacular" effect on overall student performance, it would quickly warrant abandoning the old for the new, but at what level of improvement is a major change worthwhile? Clearly, elegant as these programs were, the results of the first round of curriculum reforms did not sufficiently impress the science education community to the point of widespread adoption. It may well be, in fact, that there is little point in looking to curricular change as the *primary* means of improving student learning in science, for in one of the studies Bredderman found that program differences, on average, accounted for only 5 percent of the observed variance for all student outcomes combined, while if only process outcomes were considered, 10 percent of the variance could be accounted for.[8]

The finding that process outcomes were more prominent than others should not be surprising, since the activity-based programs analyzed by Bredderman (ESS, SAPA, and SCIS) were known to be more heavily oriented toward science process than content learning. Similarly, in an evaluation of the then newly developed high school program "Harvard Project Physics," Welch concluded that "curriculum does not seem to have much impact on student learning no matter what curriculum variations are used," noting that they rarely found a curriculum impact on students greater than 5 percent of the total variance.[9] When coupled with the fact that science usually plays such a minor role in the elementary curriculum (averaging only about 30 to 40 minutes per week) one probably should not be surprised that the curriculum reforms failed to produce dramatic effects. As Welch points out, other factors, such as student ability, time on task, and teacher ability, must have played the *major* role in how well students performed on the assessment instruments used in his studies. To these one might also add such social factors as family influences, lack of subsequent reinforcement, and (particularly at the secondary school level) peer pressures and simply "growing up."

Another finding from Bredderman's study warrants particular attention in light of our earlier remarks regarding adult literacy. Bredderman reported on three studies designed to examine former students of the new elementary activity-based science programs in the years immediately following elementary school (grades 7, 8, and 10), all finding virtually no lasting effects. Whatever advantages may have been gained during the years of exposure to activity-based science programs appear not to have been sustained in the years that followed, during which the students were subjected to conventional content-based science programs.

All three of the major programs in the first round of curriculum reform (ESS, SAPA, and SCIS) were hands-on (activity-based), process-oriented curricula, meaning that wherever possible children were encouraged to base their knowledge of nature upon direct experience, that is, upon observation and experimentation, rather than merely reading about science. As for process, the programs sought to put at least as much stress on *how* one gains and understands information about the natural world as on the information itself. Many provided kits as well as teachers' guides, and in-service training of teachers became the norm; all had evaluative phases and all were widely tested. The programs sought to encourage the learner to reason, to enlarge the child's understanding of his or her environment, and to develop the child's conceptual structure of science. In short, they were designed to produce true behavioral changes in children regarding their perceptions of science. And finally, all sought to carry the message of science as seen by the scientific community. Yet despite the innovative approaches and high quality of the programs, the overall result of this massive undertaking must be considered a disappointment, if measured solely by the current level of scientific literacy in young adults who grew up during the "alphabet soup" period. If it is true that curriculum changes alone produce only marginal effects on student performance in science, it would seem fruitless to pursue this approach to improved scientific literacy. Yet science educators persisted in looking to curriculum reform as the chief vehicle of change.

▲ A NEW ROUND OF ELEMENTARY SCHOOL CURRICULUM REFORM

Almost a quarter century after funding the first round of curriculum reform projects in elementary science, the National Science Foundation concluded that to modernize and revitalize elementary science education in the United States, it was time for a "new generation" of science programs at this level. One must conclude that this followed NSF's conviction that the outcome of the first round of reforms did not satisfy the needs of contemporary science education. Obviously, it could not be because science had changed so much at the elementary school level, but rather because alternative approaches were felt desirable. Despite their high quality (in respect to how they portrayed science), the initial reforms, we have seen, failed to produce *dramatic* results; nor did they become widely established in the schools. The reasons are fairly clear, the simple conclusion being that the bulk of the elementary school teachers were not adequately prepared or supported by their school administrators to manage these programs successfully. Most felt uncomfortable with the science or with the logistical problems of handling the activity kits, or both.[10] The summer and in-service teacher training institutes, while seemingly helpful in improving the science background of the highly motivated teacher, failed to capture the interest of the average elementary school teacher, most of whom felt then, as most teachers do today, that their own preservice education in science poorly prepared them to teach the kind of science featured in the new programs. And for the most part, local school administrators, from principals to superintendents, failed to develop the enthusiasm needed to fully support the programs in their schools.

Obviously, teachers and local administrators, but particularly teachers, are the key to successful education in our schools, whether it be in science or any other discipline. The philosophy, conceptual framework, scope, and sequence of any new curriculum designed for the elementary grades clearly are aimed primarily at the teacher (and administrators), not the student. If, for want of adequate training, the teacher is unable to fully appreciate the essential features and goals of a new science curriculum, these will not be communicated to the student. In other words, one should not expect scientific literacy to be cultivated in the elementary school by teachers who are, through no fault of their own, illiterate in science.

▲ THE "TRIAD" PROGRAM

Under a new and imaginatively conceived program (the Publisher's Initiatives Program), the NSF, with much fanfare in the late 1980s, funded a number of elementary school curriculum development projects. Each of the so-called Triad projects involved a partnership of a university, a publisher, and a school district (sometimes referred to as a "troika"), an approach that was intended to ensure the cooperation of a school district in the development and testing of the materials, and of a publisher for subsequent dissemination of the product, the latter having been noticeably lacking in the reform movement of the 1960s. At least this was the premise. It is questionable, however, whether in practice the projects turned out to be true triad arrangements. It appears that one, in fact, involved a publisher only and none involved schools to the extent of the major curriculum projects of the '60s. Since the publisher was required to commit substantial financial support to the development phase of the project, its continued interest in the outcome was assured, and at the same time the arrangement provided greater resources to the developers. On the other hand, while one of the expressed purposes of this form of educational partnership was for curriculum developers to have some influence over what publishers bring into the classroom, there remained some concern that the publishers, in seeking a marketable product, might exert a subtle influence on the developers, in effect trying to avoid radical departures from conventional publishing practice. The projects have only recently emerged from their development stage, so one can only judge from their goals and preliminary materials, rather than from long-term evaluations, how they might contribute to the overall objective of the science education community to achieve some degree of scientific literacy in students.

Beyond the novel (theoretical) partnership arrangement for these projects, the Triad programs also exhibit a number of other common features, reflecting in many cases lessons learned from the first round of curriculum reform projects. Chief among these is that all are hands-on, activity-centered, inquiry-based programs, exhibiting the widespread conviction that reading *about* science is not nearly as meaningful for elementary school children as *doing* science. All have a more practical orientation than did the first-round projects, seeking to relate science more directly to the students' everyday experiences, including a discernible trend toward technology topics, and all stress cooperative learning as an effective educational tool. A common tendency is to cover fewer concepts, but in greater depth, and to make the programs less complex for the teacher to deal with (in terms of the boxes, kits, bits and pieces that add to the logistical complexity of activity-based science programs). All the programs stress science process and also seek to integrate science with other learning—on the premise that by doing so, teachers and school administrators will find the programs more attractive and hence may increase the time devoted to science-related learning activities. And, of course, all claim to take advantage of contemporary insights in cognitive development.

One interesting characteristic of all the Triad projects was the absence of academic scientists in leadership roles. Instead, unlike the first round of reforms, which were guided mainly by scientists in association with science educators, these are directed by science educators with scientists as consultants/advisors. We have seen that the control of curriculum reform projects by scientists, rather than by science educators, had been criticized by some in the past as improperly influencing the direction taken by such programs in regard to scientific content and structure.[11] Many of the earlier programs stood out not only because of their insistence upon scientific accuracy, which obviously should be the goal of all science programs, but also, as pointed out earlier, for their attempts to present science as seen mainly through the eyes of the scientist. This leads to perhaps the most significant observation one can make about the Triad programs.

The first round of curriculum reform programs, we have already noted, were too complex for the *average* elementary school teacher, both in terms of the science involved and in the logistics of handling the equipment kits. Without fairly extensive in-service training, few teachers felt comfortable with the programs, and even with such training many still had doubts about their ability to deal with the more subtle concepts of science. And today, very few of the teachers who were trained in the first round still remain in the schools, thereby compounding the problem. The fact is that the preservice education of prospective elementary school teachers fails to adequately prepare them to teach science along with everything else they must do in the classroom; and a few weeks of in-service training, if at all effective, is hardly the ideal remedy for this well-known problem. The teacher problem appears to have been recognized by the Triad developers. The project descriptions contain such expressions as "teachers are not intimidated," "classroom-practical for teachers," "user-friendly (to the teacher) materials," "teachers feel comfortable with the program," and the like. The question is whether in seeking to make these programs "teacher-proof," the science (content and process) must be watered down to the point where the teachers fail to see science as it really is.

An obvious question is whether the Triad programs are likely to be any more successful than their predecessors in developing what might be considered a kind of scientific literacy. Here the answer, in the view of many observers, including this one, is a clear "no," for if the earlier programs were not successful in this respect, the Triad programs are even less likely to succeed. The question refers not only to whether these programs will produce a more literate adult population, which, as we have seen, did not result from the earlier programs either, but whether the subject matter of the Triad programs can be viewed as contributing directly to the scientific literacy of students, even on a contemporaneous basis. In this respect the answer must also be negative. The curricula in question do not stress the most fundamental feature of science, namely, how the laws and theories (the "big ideas") that science has evolved may be used to account for the observed facts of nature. It is wrong to assume, for example, that some understanding of the concept of energy, or even of how energy is transformed from one form to another, makes one truly literate on the subject of energy, for without an appreciation of the overarching conceptual scheme of *conservation of energy*, the students, and particularly the teachers, are left with a substantial void in their perceptions of what science is about.

In fairness, the Triad programs on the whole did not pretend to have scientific literacy as a primary goal. Those Triad projects that do claim scientific literacy as an objective use the term very loosely, that is, in the generic sense rather than in the meaning of scientific literacy that we defined earlier. Indirectly, all science education, particularly inquiry-based learning, might be said to contribute to the ultimate goal of scientific literacy, but it is a mistake to believe that young students could, on their own, in effect, synthesize the pieces of any of the Triad programs into a comprehensive view of the scientific enterprise—or even that their teachers could do so. The programs should result in some degree of science awareness, but as for scientific literacy, perhaps the best that could be hoped for is a liking for what these programs call "science" (or at least not a complete "turn-off"), and perhaps some feeling for how science goes about its business. To the extent that such outcomes *may* cause students to feel more comfortable with science later on, or perhaps even encourage some to pursue science or engineering careers, the Triad programs, if carried through as initially envisioned, will have accomplished all that could reasonably be expected of them in the face of poor teacher preparation. Some knowledgeable observers, however, have privately characterized the Triad experiment as a dismal and costly (upward of twenty million dollars) failure.

▲ SCIENCE, TECHNOLOGY, AND SOCIETY (STS) IN THE HIGH SCHOOL

A new approach to science education, Science/Technology/Society (STS), now entering its third decade, seeks to keep alive the hope that some measure of scientific literacy may yet be achieved in the educated public by reorienting the focus of science education.[12] Where in the past, preparation for responsible citizenship was often cited as a major rationale for requiring science of all students, STS sets out to formalize the societal impact of science by incorporating it directly into the science curriculum. The question is whether the STS movement, which appears to have caught the imagination of an appreciable part of the science education community, will be any more successful than past efforts to develop a meaningful—and durable—level of *scientific literacy* among our high school and college students. The prospects for this, while seemingly brighter on the surface, remain rather bleak for much the same reasons that have long plagued the scientific literacy movement itself, namely, lack of a clear-cut definition of STS and want of adequate incentive on the part of students. More than this, however, STS suffers from a terminology that seems to promise more than it can deliver in the way of science understanding (or awareness) and has attracted a number of opportunistic fringe elements, both of which make it suspect in the eyes of most scientists and many science educators.

It is instructive to examine some of the reasons why much of the science education community, and to a lesser extent the science community itself, has embraced STS education. The basic reason is that *conventional* science education has clearly failed to achieve its goal of a scientifically literate public. In the entire history of the American science education movement, we have never managed to attain so lofty a goal and, some believe, we never will. The most frequently heard argument is that the science we teach is not relevant to the students' everyday life, so that in addition to its being difficult to learn there is no compelling reason for the general student to want to learn science. The truth of this proposition is readily apparent and was discussed earlier in some detail.

Hence, one theory of STS education is that making science relevant to the students' lives may cause them to take more interest in the subject and work harder at grasping it. Another is that by making science education *socially* relevant we may also be striking a blow for better citizenship. That is, by awakening in students an awareness of societal issues that are said to be *science based* we may encourage them to take a greater interest both in science and in the societal issues, with the result that in their adult lives they may be able to play more effective roles as productive members of society. A big question, of course, is whether placing science in the context of social concerns will (a) provide the necessary incentive, and (b) also do justice to the science; that is, will it result in a reasonable level of scientific literacy from the science point of view, or only in a form of civic literacy that is more social science than natural science, for example, the "civic scientific literacy" cited by Shen.[13]

Suitable STS issues abound. Some are strictly local in nature, but most technology/society concerns are of a global nature—nuclear weapons testing, nuclear power, pollution of the air and water resulting from industrial emissions, global warming, and so forth. With some variations, such concerns are much the same in all industrialized democratic nations, even to the extent of the NIMBY (Not In My Backyard) factor. That is, a particular society may favor nuclear power on the rational grounds that it produces far less pollution than power derived from fossil fuels, or because the society lacks a natural resource of fossil fuels, yet object to it on NIMBY grounds. In fact, most siting issues, whether they be power reactors, incinerators, garbage dumps, or prisons, face the problem that local communities will accept their need —as long as they are situated elsewhere.

These are problems that more often have an economic rather than a scientific base, namely, the fear that real estate values may collapse as a result of situating what are normally considered undesirable activities within a community. Usually, any strictly technical questions involved are relatively minor and easily handled by experts if the community wishes to turn to them. Such questions include the possible effects of incinerators or dumps on local water supplies or ecologies, or the safety of power reactors, including evacuation routes in the event of catastrophic failures. But even where these turn out to be negligible risks, most people want zero risk when the potential threat is situated nearby, whereas they would be willing to accept greater risk when the source is more remote. Hence the real issues tend to be more psychological and emotional than technological, as shown by the many acronyms that have emerged, such as NIMBY, NULU (No Unwanted Land Use), or NIMTO (Not In My Term Of Office).

A central question is whether STS can take the place of conventional science for the general student. This is highly improbable, we shall see, if what is meant by this is whether the student will better understand science for being exposed to STS. Most so-called science-based social issues, we shall see, are really based in technology; hence, if anything, STS offers a route to technology education rather than science, which, if successful, would itself be a major accomplishment. We must admit that STS offers an attraction not found in conventional science courses, namely, the opportunity for direct student involvement in issues that many consider to be science related.

One of the problems of science education is that at the student level there is little or no room for personal opinion, so that students become passive participants in the learning process, asking questions but unable to contribute in a *meaningful* way to the science itself. The reason, of course, is that by the time science gets packaged into textbooks for the general student it has already attained consensus in the scientific community, and any conceptual uncertainties or errors are likely to be so subtle as to fall outside the ability of students to suggest sensible alternatives. Science texts often contain errors, of course, because scientific theories are rarely complete. Consider again the example cited earlier (chapter 3) that in the 1930s and 1940s, before the 1932 discovery of the neutron could find its way into high school textbooks, most chemistry and physics texts at this level had protons and electrons as constituting the nucleus of the atom, which accounted reasonably well for much of what was then known about nuclear properties. Would it have been reasonable to expect even a very bright high school or college student, having no prior knowledge of the field, to dispute both the text and instructor, and postulate the neutron not simply as a combination of proton and electron, but as a completely new particle in its own right? Obviously not. Unlike most other disciplines, the hallmark of science is self-correction, but modern science has grown too complex for the uninformed to play a significant role in that process. Unfortunately, by its very nature, science education tends to be authoritative. Where a secondary school student may be very good at writing original themes and short stories, or personal commentaries on some social, political, or historical subjects, that same student cannot be expected to contribute original (and useful) insights in science. One can cite many such examples, all leading to the same conclusion: controversy, debate, and personal opinion—on which the brightest students seem to thrive—by the very nature of the discipline cannot contribute to a *conventional* science curriculum, but can play a significant role in STS, as long as the debate focuses on the science/society interface and not on the nature of science itself. Thus, at least in principle, STS affords a rich opportunity to involve students more actively in what may turn out to be *the* major component of scientific literacy, namely, the impact of technology on society.

But STS has some potentially fatal flaws. First, it has a serious identity problem. STS is not yet a discipline in its own right but basically a multidisciplinary research field that attracts scholars from a variety of callings, some from the science and engineering professions but most from law, ethics, psychology, political science, philosophy, history, anthropology, and sociology. All are drawn either by social concerns or by research interests in policy matters dealing with science, government, and society. STS defies simple definition; for proof of this one need only look at the fragmentation in the field and the questions that serious research scholars in STS keep raising about the nature of the enterprise itself, questions that are echoed by many science educators as well. In universities one finds very few STS departments, graduate or undergraduate; most are interdisciplinary "programs," with the usual problems of attracting academic and financial support to nonmainstream activities.

Part of the problem of definition undoubtedly stems from its roots. World War II, and the atom bomb particularly, led many scientists to engage in political activity in the 1950s, resulting in critical examinations of the interactions of science and technology with government. This evolved during the 1960s into a broader questioning of values surrounding the impact on society of science and technology in a number of areas such as space policy, biotechnology, pollution, toxic wastes, climatic changes, and so forth. Thus the central theme of STS today, according to Kenneth Keniston, a psychologist who then headed the STS program at M.I.T., "is to try to somehow understand the complicated two-way relationship between science and technology on the one hand and the rest of culture-society-politics on the other."[14] But such a definition, understandably, sets up guidelines for a research activity, not for a school science curriculum. STS still lacks a clearly defined structure on which to build such a curriculum. Nor will it be easy to define such a structure in the foreseeable future because of the many conflicting views that surround the STS movement.

At the same symposium, sociologist Dorothy Nelkin pointed to some of the basic difficulties in reaching a common definition of STS:

> STS is still struggling for a framework and a mission. Is its purpose to promote science, advance science, and frame policy that will advance scientific and technological development? Or is it a form of criticism focused on assessing and analyzing and critiquing science and technology decisions? Is it a theoretical analysis of science as an extension of the sociology of knowledge and in effect a search for an understanding of science and society dilemmas? Or is it a policy field intended to engineer solutions to such dilemmas and to design new policies?[15]

Thus, while STS has become a serious research field for many scholars from different disciplines (mostly nonscientists) who seek a better theoretical understanding of the dynamics of the science/society interface, where precise definition of the field is not so critical, efforts to bring STS into the classroom have been fragmented. Much has been written about the goals of STS education, and several organizations have been formed to promote such education, primarily in the secondary schools, but the question remains whether the goals are really attainable, or whether STS will turn out to be just another educational fad. To sum up briefly, the advocates of STS education assume that it will better prepare students to live in our increasingly technological world by demonstrating to students in a meaningful way the interconnections between science and technology on the one hand, and society on the other. But how does one do this without a clearly defined (and workable) methodology? Turning to a now familiar example, it is easy enough to set *scientific literacy* as the goal of conventional science education. It is somewhat more difficult, we have seen, to agree upon a simple definition of scientific literacy, but let us assume that this were doable as well. Where the process breaks down, then, is in the execution; that is, how do we go about achieving the goal? With what curriculum, what instructional techniques, what evaluation instruments, do we motivate the students? These problems are compounded in STS education because it is "neither fish nor fowl" but a little of everything. Instead of scientific literacy what is being sought here is "STS literacy," which in some ways might be considered an even more demanding form of literacy. In seeking an operational definition of STS education, Richard Weirich proposed the following: "STS education includes the investigation of a societal issue which has a problem related to a scientific concept or discovery, and/or an existing or potential technology. STS education should further enable students to synthesize the various elements associated with the issue and, through the use of rational reasoning processes, reach a decision on the problem."[16]

Others go further in their specifications of STS. In setting out the competencies that should be developed through STS education, P. Rubba and R. Wiesenmayer distinguish four levels, with the last of these, the STS Action Skills Development Level, encouraging the student to take social action on STS issues through "consumerism, legal action, persuasion, physical action, and political action."[17] One wonders what is intended by "physical action," but hopefully this means no more than peaceful demonstrations.

These are by no means the elements of a modest mandate, for in describing an "STS literate" as one who is able to reach a rational decision on a technology-based societal issue by considering *all* the elements involved therein, and then acting responsibly upon such decisions, we may be expecting far more of the average student than we have any reason to believe possible. Not only does such an objective demand a relatively high level of scientific literacy, but also good measures of technological literacy, social literacy, political literacy, behavioral literacy, *and* especially good common sense. Hoping that students will achieve all this, where in the past they were not even motivated toward scientific literacy alone, seems somewhat unreasonable. Unfortunately, Weirich's definition (as is true of all others) fails to specify objectively the levels of scientific and technological literacy needed for STS literacy, or how the STS framework might encourage such literacy.

Rubba simply states that his Level 1, the STS Foundation Level, "should provide students with sufficient background knowledge, (a) of the concepts in the natural sciences and social sciences, (b) on the nature of science and technology, and (c) in the characteristic interactions among science, technology, and society, to enable them to make informed decisions on STS issues."[18]

This leads to the obvious question of whether scientific and technological literacy are prerequisite for STS literacy, or a result of it. If the former, we are back to square one; if the latter, the STS movement is on weak ground indeed, for while added social consciousness may result from such a curriculum, it is unlikely that even a modest level of scientific literacy will be an outcome as well. Indeed, it is just as likely that more students than in the past will develop a hostility toward science and technology as a result of poorly presented STS courses, for example, courses in which many of the world's ills may be perceived by students as being caused by science and technology. And there remains the obvious question of who should teach such courses—science teachers or social studies teachers, or both? Obviously, the demands upon teacher competence for STS education far exceed those for any of the individual disciplines involved. To assume that *any* high school science teacher, or worse, *any* social studies teacher, is competent to teach STS is to virtually guarantee that STS as an educational format will eventually collapse under its own weight. If finding competent science teachers is normally a problem, think how much more difficult it is to find capable, truly rounded STS teachers.

In several articles reviewing the progress of STS education during the past two decades, the historian of technology Stephen Cutcliffe concluded that on balance the STS movement was gaining ground, that the science education community had come a long way in recognizing that science and technology are inherently value-laden enterprises and that society should have some voice in determining their direction.[19] Yet in summarizing where we stood at the time in STS education, he has more questions than answers—such basic questions as what comprises an STS curriculum, where should it be taught and by whom, and what should be the proper balance between the societal and technical components of such courses. He also noted, *incorrectly*, I believe, that we have moved away from an "unthinking critical anti-technology reaction." As for his conclusion that the science education community now recognizes science and technology as being value laden, this may be more wishful thinking than factual. If correct, it can only mean that this message has been brought to teachers by social scientists who, by lumping science with technology, themselves lack a clear understanding of the nature of the scientific enterprise.

STS education has problems that go beyond questions of definition and curriculum format. In a sense all academic disciplines are established and molded by scholars in the field whose primary function is to produce future scholars. But to an even greater extent than the sciences, STS as an undergraduate or precollege educational movement suffers from being primarily a research activity. It is an educational stepchild and will remain so unless a strong STS movement emerges in our undergraduate colleges, which can happen only if all constituencies involved—educators, scholars, and students—become convinced of its educational merits. This may very well occur, but if it does it is unlikely to be in the framework of STS as presently conceived, that is, as a vehicle for science education, but rather one that deals primarily with the judgmental and confrontational aspects of technology.

Another flaw in the STS movement is the impression conveyed by its title that STS is intended as a science curriculum. In fact, science, as it is normally understood, comes off a poor third in the STS hierarchy, with the societal aspect, perhaps better characterized as a form of social studies, being its primary function. Science process skills, for example, as well as a mastery of science content, should be deemphasized in STS, according to some leaders in the field.[20] On the other hand, recognizing societal problems, particularly those having some local impact, is regarded as the main function of STS, and whatever understanding of science or technology may be needed to reach a decision on a given problem is incidental to this function. Hence, what is likely to result from an STS course is some peripheral knowledge of the science that is directly related to any societal problems that are being dealt with at the time, rather than a basic understanding of science in a way that can be applied to other societal issues. Not that there is anything wrong in this. In fact, it may well be all one can expect in the way of scientific literacy in the general public; but in that event STS cannot be considered a substitute for a *science* curriculum.

The role of STS education as a forum for evaluating (and judging) the societal effects of technology warrants further attention, for not only does this aspect of STS appeal to students, but many scholars in the field view it as the most important feature of STS. Some, we have seen, like Langdon Winner (chapter 5), unfortunately take the position that becoming literate in science is a disadvantage for one wishing to criticize the establishment. Unhappily, *unreasoned* criticism is also the stock in trade of most modern-day Luddites, and the STS movement seems to attract more than its fair share of these individuals who view STS as a convenient vehicle to attack government, big business, and ultimately science itself. Should this view of STS as being primarily a platform for criticism of science and technology prevail in the classroom, it is easy to predict that the actual science and technology components will become even more secondary and eventually disappear, leaving STS education purely an exercise in social studies. This may not be altogether ominous, for there certainly is much to be critical of in uncontrolled technology. But the education community (and the students) must then realize that STS cannot be considered a route to what is normally thought of as literacy in science and technology but only a means of illuminating societal problems that arise from the practice of modern technology—and perhaps also of pointing to action programs for easing the public mind on such issues. Many social reconstructionists evidently believe that this is all the science and technology literacy one needs, namely, knowing how to react to perceived threats stemming from the use of technology. But one can hardly equate such kneejerk responses with any sensible degree of literacy.

Not that conventional science courses do so well for the general student. We know that in some ways, for example, in achieving scientific literacy, they fail their purpose. But unfortunately STS has the potential for doing worse by conveying a misleading impression of science. First of all, it is a mistake to believe that STS can turn students on to science as such, for most societal issues that are said to be science based are really based in technology. The list of societal issues that are truly *science based* is very short indeed, encompassing mainly such questions as federal funding for research (e.g., space probes, the Supercollider, the Human Genome project), or whether certain types of scientific research should be discouraged, or even prevented (e.g., animal experimentation, genetic engineering, human gene transplants). Hence if any literacy is to be gained from STS it is more likely to be some form of technological literacy rather than scientific literacy. Not that this would be undesirable, for as I have pointed out, if widespread technological literacy could, in fact, result from STS education, plus, one would hope, an awareness of the partnership between science and technology, STS will have accomplished far more than all past efforts to provide students with some *durable* understanding of the overall scientific enterprise.

A potential barrier to widespread acceptance of STS is that while in theory STS offers the potential of becoming a serious contender in the science curriculum contest, some science educators are concerned lest it become simply a forum for social criticism of science and technology, which much of it already is. We have seen that by its very nature, STS seems to attract certain fringe groups of technology critics, neo-Luddites, anti-science cultures, and those who regard it as a convenient vehicle to lay many of the ills of modern civilization at the doorsteps of science and technology. Should such groups succeed in distorting STS to their own design, or should STS become dominated by the social science community (which, while obviously having an important role to play in the STS concept, nevertheless has a different agenda than the science education community), the STS movement, at least as it concerns the school *science* curriculum, could very well self-destruct.[21]

One can predict that all such efforts to put a "social spin" on the practice of science will fail, just as they did in the Soviet Union during the Lenin-Stalin era, when prospects were brightest for bringing science into conformity with Marxist philosophy. Reputable Soviet scientists simply kept their distance from the Bolsheviks intent on recasting science. Then reason prevailed, and the need for competent, high-level science and technology in the USSR ultimately resulted in casting out revisionist efforts to introduce a "people's science."[22]

Senseless though the science counterculture movement may seem, and fail though it will, it nevertheless poses a problem for science education, particularly for STS, where it unfortunately makes the greatest impact. Strangely, part of the problem is traceable to a confusion in terms, to the meaning of "constructivism" as applied by the counterculturists to the *practice* of science, and its use by some science educators to characterize a new attempt to restructure several learning theories under a single unifying theme.

The constructivist model of learning, which lately has caught the fancy of many cognitive scientists and science educators,[23] is being mistakenly carried over by some to mean that what should be taught is the constructivist view of science itself, with its fanatic opposition to the concepts of objective evidence, universal laws of nature, and value-free science—a complete shift from the rational to the irrational. Not that the practice of science is likely to change under pressure from the postmodernists, particularly as they have been notably unsuccessful in showing that their free-association methods of seeking to understand nature lead one to any useful knowledge. They are quick to denounce the principles underlying the modern practice of science (which, on the whole, have proved successful) but fail to propose any practical alternatives. Working scientists will continue doing their science in ways they have found to be most productive in terms of getting results that stand the test of time (and of their peers), even though such practice may offend the postmodernists.

The danger, however, is that, in applying the constructivist learning model to science *and* coupling it with the social-constructivist view of science itself, STS teachers may give students a badly distorted view of science. One sees this occurring in some of the recent literature directed at STS educators. B. Reeves and C. Ney, for example, distinguish between constructivist understandings about science and what they call (pejoratively) the *"naïve* positivist's" understanding, meaning what most scientists and science educators currently take science to be.[24] Needless to say, given their initial thesis that human values *must* be incorporated into the study and practice of science, conventional science does not come off well in the comparison.

To cite a more pernicious example, the first draft of a position paper on science education standards for grades K–12, issued in 1993 by the National Committee on Science Education Standards and Assessment of the National Research Council, included a statement that the standards would reflect the "postmodernist view of science," which "questions the objectivity of observation and the truth of scientific knowledge," obviously just the opposite of the positivist view held by most of the scientific community.[25] Fortunately, this attempt to insinuate postmodernist views of science into national standards (and hence into future curricula) was caught in time to be eliminated from the committee's final report. But such attempts will undoubtedly continue, and the STS community must be on constant guard against them, lest STS become just another experiment in social studies.

Judging by the extensive literature in the field, the STS movement is growing, mainly through the infusion of STS units into conventional high school science courses. The number of dedicated STS courses remains very small, as might be expected, possibly because so much of high school science is tightly prescribed, or perhaps because teachers are not yet convinced of this approach. In any event, it is much too early to tell whether STS education can provide the magic formula even for enhanced social literacy in science, let alone literacy in science. My own view is that, without showing in convincing fashion *how* students are to acquire the knowledge of science and technology requisite to *personal* decision-making on science-based societal issues, the movement will fail its primary purpose.

▲ LESS IS MORE? SCIENCE FOR ALL AMERICANS: PROJECT 2061

The complaint heard most often about contemporary science education is that science courses for the general student try to do too much. They seek to cover virtually everything that is considered basic to the subject and can be dealt with at a given grade level. Curriculum developers are generally reluctant to leave out topics that reviewers might consider essential, particularly at the high school level, where the curricula are strongly influenced by the subject matter of the corresponding first-year college courses. Despite repeated calls over the past three or four decades for more depth and less breadth, little has changed in the content of high school science courses except to keep pace with new developments in the field, particularly as these are introduced into the college curriculum. Of all the ongoing curriculum reforms in science, the AAAS initiative known as Project 2061 is the most outspoken on the premise that requiring less of students in the way of science concepts and factual information may result in better science teaching.[26]

The initial phase of the project, now completed, focused on the substance of a new curriculum, on setting down the specific knowledge, skills, and attitudes in science that one should like students to acquire by the time they complete high school. Panels of scientists and educators, with no initial constraints, were charged with determining what they believed to be *essential* for all citizens to know in a scientifically literate society—not the details, only the basics—particularly the interconnections among the various disciplines, covering science, mathematics, technology, and social science. No attempt was made to define scientific literacy directly; instead, the panels focused on questions that may be said to constitute scientific literacy, such as, "What is the Scientific Endeavor," "What is Scientific Thinking," "What is the Scientific View of the World," etc.

The product is, in fact, a blueprint for future curriculum developers, outlining for them the topics and subject matter in the different areas of science that groups of experts in those areas felt were basic to their fields. It is a masterpiece of apparent conciseness, serving perforce as the AAAS "definition" of scientific literacy. The real tests of its value, however, will be (a) the ability of curriculum developers to transform these guidelines into *effective* student-oriented course designs at the K–12 level, and (b) the lasting quality of such courses. Persuading scientists to pare down their disciplines to bare essentials is no mean feat, but it is far easier for scientists and educators to agree on what everyone should know about science than on how to persuade students to want to learn it and then to retain it during adulthood. The magic motivation formula remains to be discovered.

Project 2061 makes no pretense that what it has developed to this point, while presumably satisfying to the scientific community, is feasible as far as students and teachers are concerned; in fact, feasibility was not considered during the initial phase, the main goal being to set down the essentials of science. The final results will not be known for many years, for the second phase is a period of curriculum development, and the final and longest phase will involve large-scale testing of a number of alternative curricula developed according to the guidelines set down by Project 2061. It could be decades before a final verdict on the merits of Project 2061 can be reached. However, two concerns might be voiced at this point.

The first, alluded to above, is the question of motivation— whether, in fact, given *any* new curriculum, however well designed in terms of content, students can be turned on to science to the extent of becoming truly literate in it. If scientific literacy means knowing most of the "basics" established by Project 2061, it is doomed to fail, for at no time in the entire history of U.S. public school education has even this much knowledge of science been expected, or realized, of high school graduates. The outward appearance that Project 2061 has managed to prune down the content of all of science and technology is somewhat deceptive, for when curriculum developers start fleshing out the major topics presented, it is likely that the science curriculum will grow to the same unmanageable proportions we presently find in our textbooks.

The other concern is more significant. We have seen that science literacy in school is no guarantee of such literacy as an adult, where it is far more important in the sense of contributing to the good of society. Project 2061 fails to take this problem into account, ending its responsibility, in effect, upon high school graduation. But there is far more to scientific literacy than this, as we have seen, and it is perhaps unfortunate that the AAAS has not set its sights beyond this point.

The ambitions of Project 2061 actually go beyond merely specifying what is important for students to know about science and technology. The Project also seeks to establish Benchmarks for Science

Literacy. These, in effect, specify, according to the AAAS, "what all students should know or be able to do in science, mathematics and technology by the end of grades 2, 5, 8, and 12."[27] Obviously, such benchmarks will establish further guidelines for curriculum developers and could, in the end, intentionally or not, form the basis of a national testing and evaluation program. In the eyes of those who believe in nationwide educational standards, Project 2061 must seem the ideal solution, assuming it works, of course. To others, it will seem too prescriptive. In either event, Project 2061 will be a boon to many developers, most of whom are likely to take it as *the* gospel, while others will understand that while it may be an authoritative guide to what is considered important in science by its practitioners, it is not necessarily the last word. In this sense it will serve a very valuable purpose. Unfortunately, curricula alone, however good they may be, cannot remove the main barriers to universal literacy in science.

▲ EMULATING THE EUROPEAN PATTERN: MORE IS BETTER?

At the opposite pole to Project 2061 is the latest National Science Teachers Association (NSTA) approach to secondary school science education (grades 7–12), one that might be characterized as "more is better."[28] The Scope, Sequence and Coordination (SS&C) program of the NSTA, said to be the largest science education reform effort since the post-Sputnik days of the 1960s, seeks to replace the traditional "layer cake" design of U.S. science education, as it is sometimes called, with vertical structures resembling those employed in the U.K., Europe, the former USSR, and parts of the Far East. The proposed changes are substantial. Typical science sequences in U.S. high schools consist of discrete (usually one year) courses, primarily in biology, earth science, chemistry, and physics. These are taken sequentially, most often in the order illustrated. There is little if any coordination among them, and the vast majority of students (namely, the nonscience students) are required to take only one or at most two of the courses; hence the layer cake designation. In contrast, most students in the countries that NSTA would emulate are exposed to all the main science disciplines concurrently, every year over periods ranging from four to six years, and, at least in principle, with some coordination of the disciplines. NSTA points out that where the present high school sequence offers a total of 540 hours of science over three years, with most students taking less, the USSR provided more than twice this exposure over a period of six years for practically all of its students. Overall, the proposed SS&C plan, if fully implemented, would require that all students devote a total of some 1,500 hours to science in the secondary school (grades 7–12), about double the normal practice today.

Simply calling this an ambitious plan would not be doing it justice; SS&C is a *massive* undertaking that faces a multitude of problems having little to do with science education per se. Not the least of these is where the additional time will be found. It must come from a combination of decreased free periods, increased school day and/or school year, or by eliminating or decreasing the time spent on other subjects or school activities. While the first options are probably much to be desired, the last will undoubtedly face a great deal of opposition in some quarters, as well it should. Then there is the question of whether the premises underlying the proposed reform are valid, and here one can take serious issue with the SS&C assumptions.

What are the basic objectives of such vast reform? Is it intended to bring more students into the science and engineering pipeline (which currently is a highly questionable premise), or simply to upgrade the general literacy in science of all students? Presumably the NSTA and the NSF felt encouraged in their strategy to force more science upon all students by the recent movement in the United Kingdom, where after many years of political debate, motivated by the fear that its industrial strength was deteriorating, science education was made a dominant feature of their National Curriculum, with all students being required to take science throughout their years of compulsory education, roughly ages five to sixteen (in some cases, age eighteen).

The new British core curriculum consists of English, mathematics, and science, with science making up about 20 percent of the total curriculum. In fact, science now outweighs mathematics and general "literacy" in the National Curriculum, thus placing it above the traditional 3 R's in importance. Most scientists and science educators in the U.K. naturally applaud the new curriculum, obviously believing it to be for the good of their professions, if not also for the nation. In commenting on the wisdom of this move, the chemist and past president of the Royal Society of Great Britain, Sir George Porter, is reported to have asked rhetorically: "Should we not force science down the throats of those that have no taste for it? Is it not our duty to drag them into the twentieth century?" His answer, of course, was positive,[29] evoking memories of Herbert Spencer some 150 years earlier.[30] But now that the new curriculum has been adopted, more moderate voices are being heard. In a detailed, perceptive article, the British science educator Bryan Chapman challenged the new curriculum on all conceivable grounds—educational, political, economic, industrial, social, and philosophical, much the same as we have delineated them here. He concluded that experience will show it to be little more than politically motivated folly.[31] Indeed, there are some in the U.K. who believe that if the new curriculum is even moderately successful in turning more students to science, the supply of trained scientists and engineers will far exceed the demand, unemployment will result, and many will seek employment abroad. Unfortunately, the full impact of major educational reforms do not become apparent for a generation or more, so that it is too much to expect that the British experiment, or the AAAS and NSTA curriculum reforms in the United States, will themselves be subject to critical reexamination and further reform in only a few years.

Fortunately, some would say, there is a major difference between the new National Curriculum experiment in the U.K. and the U.S. curriculum reforms described above, namely, that the latter are not carved in stone on a national basis, despite the ultimate objectives of their promoters, thereby making it possible to experiment regionally before any widespread adoption of a new curriculum. This results from control of U.S. education by the states rather than by a centralized government agency. In this instance, we have a system which, while often criticized for lack of uniform national standards, serves as a useful buffer to any attempts to establish a new (untested) curriculum on a nationwide basis.

England was not alone among highly industrialized countries to restructure its science and mathematics education in recent years. The Soviet Union introduced a new curriculum in the 1986–87 school year, shortly before Gorbachev's *glasnost* and *perestroika* policies, when Soviet society and political structure were virtually unchanged from those of the Stalin era. As with the new British National Curriculum, the Soviet curriculum was made prescriptive from age six (1st grade) through age seventeen (11th grade). According to a study by C. Ailes and F. Rushing, "One of the most distinctive features of the Soviet educational system at the precollege level is its attempt to provide uniform science and mathematics instruction for all students up to the completion of the secondary school process. The national curriculum is laid out in a precise manner with each student at various levels (primary through secondary) progressing through it at the same rate."[32]

Problems of implementation aside, by dealing mainly with the question of how science instruction might be better organized and delivered rather than its specific content, SS&C represents the *only* major departure from past efforts to improve science learning in our schools. Until SS&C, the usual route was through curriculum improvement and, to a limited extent, improved teacher preparation. My reservations as to the long-term effects of curriculum materials reform alone have been made amply clear. Here at last is a radically different kind of experiment. One might challenge its premise, but assuming that the evaluation phase is well designed there can be no doubt of its enormous potential value to educational research in the United States.

▲ THE NEW LIBERAL ARTS

Practically all efforts to improve science education on a national scale, starting with the initial programs of the 1960s, have been at the precollege level. This is true as well of the current round of federally supported programs. The full reasons for this would make an interesting dissertation, but central to these was the widespread (mistaken) belief on the part of college science faculties that the problem was really not theirs but belonged in the high schools and below, and should be solved by the schools of education. While many university science faculty participated in, and even initiated and directed some of the precollege curriculum projects of the last three decades, the same attention was not given to a critical analysis of the college science requirements for (nonscience) undergraduate students, or to efforts aimed at restructuring such undergraduate education in the framework of scientific literacy. Many new courses were introduced at the local college level, most having titles designed to attract the nonscience student, as we have seen, but no widespread organized movement along these lines ever resulted.

One of the most innovative programs in contemporary college science education, perhaps wishfully labeled "The New Liberal Arts," had two principal goals: increasing the attention paid (1) to quantitative reasoning (special topics in mathematics), and (2) to technology in the undergraduate curriculum of liberal arts colleges. The program, recently ended, was founded and supported by the Alfred P. Sloan Foundation for about a decade through a number of workshops, clinics, seminars, and substantial grants to dozens of colleges and universities to develop new course material aimed at the stated goals. The premise (and title) of the program reflects the feeling of many educators, mostly in science and engineering, that despite the growing influence of these fields on modern society, a liberal college education is still largely a classical education, and that these two areas, quantitative reasoning (numeracy) and technology, are among the major deficits of liberal arts graduates. One might quarrel with these as being *the* major deficits, compared, for example, with an understanding of the nature of science, but there can be no question of their importance. A number of excellent monographs, books, and extended syllabi on specific course designs were produced by the program during the ten years of its tenure, all providing a wealth of curriculum ideas relating to the two principal themes.[33] Overall, this program deserves the highest marks for its conception, operation, and end products, and could well serve as a paradigm of future curriculum development at the college level.[34]

A final note is in order on all of the curriculum reform efforts now underway, both in the United States and in other advanced nations. None seem to address the important but very difficult question of durability or "staying power"—of how, if at all, these curriculum reforms will bear on the scientific literacy of adults, indeed, even of those adults who as students performed very well in science, regardless of the kind of curriculum then in vogue. I have pointed out time and again, yet cannot emphasize it too much, that in a practical societal sense it is far more important to have literate adults than literate students, and while the two are not necessarily mutually exclusive, there is no evidence of a close link between them. In fact, in those countries which have traditionally employed the vertical "spiral" approach to science education, such as that currently advocated by the NSTA and made compulsory for all students in the new U.K. National Curriculum, the adult populations appear to be no more knowledgeable about science than our own. This is a reality that most curriculum reformers seem unwilling to recognize as an essential component of the overall problem.

CHAPTER 7

The Future of "Scientific Literacy"

The Shape of Things to Come
—H. G. Wells (Book title, 1933)

▲ No one can accurately predict what general education in science will be like a century from now, but because major educational changes tend to occur slowly it is possible to project from current trends what is likely to happen over the next generation or two. We are concerned here not so much with the science-bound students but with the so-called general students, who comprise the vast majority of students in our high schools and colleges. It is easy enough to forecast the future of science education for those students who wish to become professional scientists or engineers. If anything, their training will become more specialized as the major science disciplines continue to fragment into subspecialties and the growth of scientific knowledge maintains its incredible pace. Generalists will become scarcer, and the main problems of curriculum reform will be to discard older subject matter in favor of new, or to extend the training period of science-bound students, or both. Aided by federal funding, probably the number of specialized science high schools will increase, both at the state and local levels, particularly if a time is reached where the projected growth in science and engineering takes a sharp upturn.

More will be said later about the science-bound student, but what is in store for the 90 percent or more who are not particularly interested in science? What does the future hold for the scientific literacy "movement," which is supposed to stimulate the minds of this large group of students and propel them into the twenty-first century with a far better grasp of the nature of science than that held by their parents and all earlier generations? At this point, science education is riding the crest of a new wave of congressional support, which is likely to result in increased budget allocations to NSF and the U.S. Department of Education over the next few years in support of efforts to improve this area of education.

Scientific literacy has become too good a public relations "prop" for the science education community to abandon at this point. It has served the community well as the rationale for increased support for science education. Eventually, however, as it fails to produce concrete results and as the folly of resorting to such a poorly understood label becomes obvious to most educators, it will prove to be a debit rather than an asset and will be abandoned in favor of some more meaningful goals. The goal of "scientific literacy" has become almost synonymous with today's science teaching. Never mind that its meaning remains vague and that the methods of achieving whatever one takes it to mean are still unknown; ask any present-day curriculum developer the purpose of science education for the general student and the answer is almost certain to include scientific literacy as a major objective, if not the primary one.

Over the next few years we will see a continuing flood of books and articles deploring the sad state of scientific literacy in America, many offering formulas for reversing this "trend," as it is often called. Actually, it is hardly a trend but rather business as usual, for we have already seen that never in U.S. history has there been a time when the public at large, or even its highly educated segment, could be considered literate in science. The solutions offered will invariably require large expenditures for curriculum development, testing, and implementation. None of the credible remedies will be short term, for we know that major educational reforms cannot be fully evaluated in less than a decade, and most take two or three decades for a final verdict to be reached. There is more to the problem than designing a new curriculum, however promising a given approach may seem in principle. Who will be equipped to teach it? And, as long as textbooks remain the chief currency of information transfer in the schools—which should be a long while if one includes both print and video media—who will publish new and innovative curriculum materials?

For obvious commercial reasons publishers cannot afford to be innovators; they generally wait until a curriculum reform shows promise of adoption, or other sources of support, before committing extensive resources to it. Yet without commercial publishers to promote and distribute new curriculum materials, their dissemination will languish and worthwhile programs may not get the attention they deserve. Instead, the future science classroom will revolve about those curriculum reforms that have been most strongly supported by the major funding agencies, which generally means those that promise the most dramatic changes from previous programs, notwithstanding that they may be impractical or ineffective. Meanwhile, popular books will continue to assert that science is simple and proceed to address the limited number of readers who are attracted by titles or reviews. Both the books and reviews are often written by science literates, but the uninformed reader, while possibly finding the subject matter interesting and informative, rarely ends up with the much hoped-for appreciation of the true nature of science simply from having read such books. One such example is a recently published book by R. M. Hazen and J. Trefil.[1] Trefil first came onto the science literacy scene a few years ago by coauthoring, with Hirsch and Kett, the dictionary of cultural literacy discussed earlier.[2]

Hazen and Trefil start from the reasonable premise that scientific literacy comprises the knowledge one needs to understand public issues. We have already pointed to a basic flaw in this premise, namely, a lack of precision in what should be meant by "understanding" (chapter 4), but assuming the common interpretation of this term one can find no serious fault with the premise. The authors then discuss a number of "great ideas" in science to help the reader achieve this end, and here the book does miss its mark. The great-ideas approach to scientific literacy is by no means new and suffers from the same syndrome that afflicts most contemporary science education designed for the general student, namely, an emphasis on science content rather than on the nature of science and the scientific enterprise, and the risk that not all scientists would agree on the same list of great ideas.

Obviously, Hazen and Trefil do not suggest that merely learning these great ideas by rote, which is what most students do, constitutes scientific literacy. But nor do they take into account the poor staying power of scientific concepts, great or small, when taught in the abstract, that is, without a solid grounding in the nature of science itself. Unless the educated adult population of our nation retains something of what they were taught in science—and the evidence is clear that up till now they have not—the notion that we are teaching for scientific literacy, however one may define this slippery term, in that segment of society where such literacy might really count, is pure fantasy. While their book does not make the point explicitly, the authors apparently believe (judging from other of their writings) that science *appreciation* may be a more suitable goal than scientific literacy for the general student, a position that I readily accept (see chapter 8).

▲   CHALLENGING THE PREMISE

The only reasonable conclusion one can draw from the massive evidence accumulated over nearly a century of experience in the U.S. science education movement, and particularly during the past half century of seemingly endless "reforms" in science education, is that there is no consensus on what "scientific literacy" means or should mean. Instead, everyone involved with science education appears to have a vague, ill-defined notion of what it should mean, ranging from the simplistic view that any exposure to science contributes something to the state of mind called "scientific literacy," to the equally naïve view that scientific literacy means being able to think like a scientist.

If science education has learned anything from the past, it is what does *not* work in terms of developing in students a rational and *lasting* impression of science and its interplay with society. The truth is that despite the reluctance of teachers and science educators to accept the fact that their efforts may have been in vain, nothing we have tried works for the majority of students, except perhaps in the elementary grades where the criteria for success are obviously quite different. For the simplest proof of this one need only consider the current intense activity surrounding science education—a clear admission that even a modest solution is yet to be found.

At this point, science education has only two options, other than the possibility that if nothing is done the problem may go away of its own accord because people will tire of it. The first is to take our dismal lack of success to date as proof of the impossibility of achieving universal scientific literacy and give up the notion of compulsory science education for all, leaving the choice to the individual student. Or second, we can assume that the two mainstays of educational improvement, namely, curriculum reform and improved teacher training, have not yet been attacked aggressively enough in science to cause a marked difference, and look for *major* changes in these factors, instead of the relatively small increments we have witnessed thus far, to produce positive results. As a practical matter, it is easy to see which will prevail in the near term. One cannot imagine the science education establishment willingly giving up the strong toehold it has established in compulsory precollege and college education. And given the endless optimism of the education community, nor can one see many science educators (or granting agencies) willing to concede that there is no real hope of achieving the elusive scientific literacy.

Hence, we must accept the reality that curriculum reform efforts will go on and ask ourselves what kind of reforms might produce a *useful* result. The sensible answer here, as I have tried repeatedly to stress, is to make a fresh start, to begin by challenging the premise that scientific literacy (meaning, as commonly understood, a grasp of science method *and* content) is the sine qua non of any educated person at the end of the twentieth century, and to reassess the purpose of science education for the general student. Starting with a clean slate we may find that solving a subset of the larger problem will not only be more useful to our society, but may actually be achievable. By examining the specific failings in public understanding that play an important role in fashioning public attitudes toward science, as well as those shortcomings that somehow impact the individual's own safety or quality of life, it is more likely that a meaningful guide to general education in science will result. There are no easy solutions to the overall "problem," if indeed one regards scientific illiteracy as a pressing social problem. Many people obviously do not, considering the amount of lip service that is accorded the issue in some educational circles. Nonetheless, even if the problems of scientific literacy are not readily solvable, some positive steps toward a reconciliation of science and society may still be possible.

▲ MANDATING SCIENTIFIC LITERACY?

Clearly, for any new science curriculum to gain widespread acceptance it must, in the end, also satisfy some perceived need of the students rather than only the science education community or adult society. What might constitute such incentives? First, if all colleges and universities were to make known, even informally, that excellent *performance,* not merely completion of courses, in precollege science and mathematics would be considered very high on their admission profile, college-bound students seeking admission to institutions of their choice would have the incentive to do well in these subject areas, just as most students now regard as important performance on the two dominant college admission tests, the Scholastic Aptitude Test (SAT) and the American College Testing Program (ACT), each of which is taken by almost one million high school students every year (roughly one-third of all high school graduates). However questionable these tests may be as accurate predictors of college performance, they are nevertheless regarded by all but the very brightest students as a major (and traumatic) hurdle that must be overcome, one that becomes uppermost in their minds as they enter their senior year in high school.[3] If these were replaced by a national *achievement* test as a measure of probable college performance, as many now urge,[4] serious students would no doubt seek to prepare themselves for such examinations as they now do for the SATs and ACTs.

In the pre–World War II period, before there were SATs or other standardized tests as guides to future performance and college admissions officers relied mainly on high school records and teacher recommendations, few of the leading American colleges listed science as a formal admission requirement. Yet it was well known among high school students that in practice these colleges, along with those that actually specified science and mathematics credits for admission, also looked to performance in science as a measure of probable success in college. As a result those students who were college-bound elected as much high school science as they could manage, bearing in mind that good grades were also prerequisite to college admission. Thus the incentive to do well in high school science was there, unlike now, when our state and community university systems virtually guarantee every high school graduate admission to a college, although not necessarily to the student's college of choice. This is not to suggest that the incentive of preparing for college admission is strong enough for students to attain a *meaningful* literacy in science and mathematics, but only that the evidence from all other industrialized nations, which do not offer "universal" higher education, is that students enter their colleges and universities far better prepared in science than do their counterparts in the United States.

Another obvious way of "forcing" science on students is through the job market. If business and industry were to make known in unequivocal terms that the modern work force *must* be better prepared in science and mathematics than it has been in the past, implying that job opportunities not only in science but in other fields as well would depend upon such knowledge, students (and teachers) would surely respond to meet the employment conditions. However, the prospect of the business-industry segment mounting a *credible* appeal, one that could be satisfied only by formal education rather than through the usual in-house or other specialized training, is unlikely for the simple reason that the evidence supporting such a pronouncement would hardly be convincing without corresponding incentives in the way of attractive employment opportunities, including top salaries.

Still another and equally unlikely method is the suggestion that high schools and colleges require that students demonstrate proficiency in science and mathematics as a condition for awarding their degrees. Such a "threat" might be effective in turning a few more students toward science (and very likely many more away from science), but it is unrealistic for several reasons. One is the expected resistance from the humanities and social sciences to the notion of according science and mathematics so central a role in the school curriculum. Another is the fact that some students simply have no taste whatever for science and/or mathematics, so that unless the proficiency level required of them is so minimal as to be almost meaningless they could never graduate. And finally, there is the familiar question of whether "cramming" for required examinations means much in the overall picture of scientific literacy—or any form of literacy, for that matter.

Finally, if the executive branch were to declare officially that for national economic or defense reasons, general literacy in science and mathematics is an educational imperative, and urge the states to adopt a national curriculum in this area, *and* also back up this position with significant financial support, the pattern of science education and the stature of the science and engineering professions could change almost overnight. The fact that this has not occurred, not even during the emergency climate of World War II, when many scientists and engineers were mobilized for the defense effort, must be taken to mean that the administration has never felt the need for such drastic measures.

In a sense, this has been done in the United Kingdom with their new National Curriculum, which, as outlined earlier, elevates science and mathematics into dominant positions throughout the entire school curriculum (at least to age sixteen). This was achieved by the Education Ministry, urged on at least partly by the business/industry community, to counter what was feared to be a declining British role in world industrial markets, particularly in light of its slowly weakening manufacturing base and its need to jockey for a stronger position in the European Community. Whether the move will succeed in elevating British industry to a more competitive level by raising its industrial output relative to other Common Market countries remains to be seen. It cannot be known with any certainty for at least a decade or two, this being the typical assessment period, as we know, for major educational reforms. It should prove an interesting experiment, involving educational, social, and economic factors, one that has near-ideal conditions for measuring results: a relatively small, self-contained nation having a single national curriculum, a negative trade balance, and well-defined methods of measuring its industrial and economic growth. A troublesome element of the experiment will be to take into account all possible external factors, such as world conditions that may affect the

British economy yet cannot be associated with any educational reforms.

We have already pointed out a possible major negative result of this educational experiment, but it bears repeating because of the likelihood that our own initiatives for increasing the science and engineering manpower pools may have the same end result, namely, a surplus of individuals trained in these fields who must then seek employment elsewhere. Fortunately, this would be a transient effect, whether in the U.S. or the U.K., for the supply would soon normalize as word filtered back to students that the demand had been overestimated. It is doubtful that a national curriculum will ever be mandated in the United States, but it is likely that some form of national standards and performance testing will be adopted to spur educational improvement generally. Many educators would applaud such a move, but it would not win unanimous approval on the ground that simply encouraging competitiveness is not the answer to quality education. Probably a better solution would be to advocate a consensus among the states on *minimum* competencies, particularly in science and mathematics, that must be demonstrated for a high school diploma. We shall see later that the Clinton administration has actually started to move in this direction.

When former president George Bush announced that he wished to be known as the "education president," he instituted a drive toward national testing (viewed as a means of forcing local school boards to upgrade the performance of their students), but he unfortunately failed to support the move with a substantial infusion of federal funds. And without federal support it is doubtful that much can ever be accomplished in the education area. Moreover, the executive branch did not target science explicitly, except to state that it should be the world's best. On the other hand, both the NSF and the Congress believe that teacher enhancement and preparation in science deserve the highest national priority, and they seem prepared to support such a thrust, which could have an enormous impact on the future quality of our science education. It appears that the Clinton administration will continue efforts to revamp American schools, but will reject the Bush market-oriented approach in favor of a national education policy directed somehow from Washington. However, it appears that certain features of the Bush plan, such as voluntary (age-specific) national curriculum standards and national testing, will be retained. National standards and testing remain a controversial issue, of course, one that is bound to engage the education community in a continuing debate for many years to come. But in the end, at least for science and math, it is highly probable that some form of voluntary standards will be adopted.

Thus, while specific actions by the business/industry community and/or government might force students to demonstrate proficiency in science and mathematics, they could sensibly apply only to limited areas of these broad disciplines, namely, to technology and to simple quantitative reasoning. They would apply to technology because, if there is any validity at all to the argument that students must be prepared to live and work in an "age of science," we know this means technology rather than science, and certainly not the formal coursework that is designed primarily for science-bound students. Similarly, the capacity for quantitative reasoning obviously is not gained by the general student through conventional courses in geometry or algebra, or through ill-fated curriculum designs like the "New Math" introduced in the 1950s and virtually abandoned some twenty years later. Hence a totally different approach is needed if such reasoning is the desired outcome of formal mathematics education, just as a different approach to science education is needed if public understanding of science is to be its objective. Modest changes will not suffice; instead, as will be discussed later, a "paradigm leap" will be required in our approach to science and math education. As we have seen, however, barring a clear national emergency that might dictate extreme measures by business, industry, and government, including the huge financial commitments needed to underwrite so massive an undertaking as converting to a true technology-based society, it is unlikely to occur in the foreseeable future. Indeed, probably most Americans hope it never will.

▲ WHAT MAKES SCIENCE RELEVANT?

This leaves as perhaps the major incentive for general students to take a serious interest in their high school and college science courses the feeling that science is, to use a much overworked term, *relevant* to them personally. What does relevance actually mean in science education, or in education generally, for that matter? To some educators it means dealing mainly with subject matter that "rings a bell," so to speak, topics that evoke images of familiar everyday things or occurrences—simple questions such as, "Why is the sky blue on a sunny day?" or "Why is the sky black on a moonless night?" or "What causes the seasons?" or "Why do the metal parts of my chair feel colder than the wood or the fabric?" or "Why is an astronaut weightless in orbit?" or "Why does it reduce the average waiting time to have a single queue feeding several tellers in a bank than a separate queue for each teller?" One can easily produce a long list of questions of this kind which one would think should gain the interest of most students; at least they cannot be labeled abstract concepts. But it does not necessarily follow that because the subject matter is commonplace students are more than casually interested in reasoning through the answers. Elementary school children, yes; their curiosity is such that almost anything about nature or science captures their interest. But not so at the high school and college levels. Here, much depends on how painless is the process of acquiring and understanding the answers. The foregoing questions, for example, although dealing with ordinary, commonplace phenomena, are not all as simple as they may seem. If a great deal is required by way of reasoning or analysis, as it is for some of these questions, many upper-level general students quickly lose interest. They much prefer simple factual answers without the need to follow the reasoning involved.

Other educators see relevancy in a different light. To most sociologists it means dealing with social issues having a science or technology base, as one might find in a typical STS curriculum. In fact, STS developed as a component of science education on just this premise, that is, as a means of getting students more interested in science by showing its relevance to their communities, if not to them individually. The problem with this is that not all students are interested in issues that affect society generally—such as global warming, nuclear power, or the environment—yet are remote from their personal experiences. For instance, it is not easy to get a class excited about the issues surrounding a proposed garbage dump or power reactor if these are not being sited in or near that particular community and hence do not involve the local media. And even when students do evidence some interest in such issues, their link to science or technology often is not very strong. That is to say, one can deal with most such issues on a societal level with little understanding of the science or technology involved, and with no appreciation whatever for the nature of the scientific enterprise. After all, this is the level at which sociologists and political scientists normally deal with such societal issues, so it should not be surprising that whatever science or technology may be involved in them can be treated superficially. Science and technology generally enter the picture in terms of technical solutions that may be offered to remedy a particular problem, rather than in understanding the societal impact of the problem.

For example, must one understand the detailed mechanism of global warming to conclude that steps should be taken to reduce the use of fossil fuels (CO_2 emitters) and other greenhouse-type emitters? Not if one starts from the premise, which is presently based more on theory than experiment, that (a) global warming is definitely occurring, and (b) it will be deleterious to *all* society, which appears to be the position taken by most environmentalists, who have already made up their minds about the issue even though *unequivocal* scientific evidence for it is still lacking. Actually, the jury is still out on the first of these as far as convincing evidence is concerned. On the second question, if significant warming were to occur, it is just as likely that some areas of the Earth (e.g., agricultural areas) would benefit from it as that others would suffer. This is unlike the issue of ozone depletion, which is believed to be caused mainly by the release of industrially produced chlorofluorocarbons into the atmosphere. It is relatively easy to get international agreement on limiting the use of chlorofluorocarbons, largely because such use is not considered essential.

I cite these examples because, as is the case with many such environmental issues, if the effects are real, one does not require much science to understand either their causes or obvious solutions. To determine whether they *are* real, however, does require one to understand how "scientific evidence" should be interpreted. Here, as in most situations involving the more subtle effects of technology, it is too much to expect the average educated person, even one who may be regarded as reasonably literate in science and technology, to make this interpretation on his or her own personal knowledge. More to the point, perhaps, is whether solutions accepted on unsupported scientific grounds will be based more on political or economic factors, or on international relations, than on purely environmental considerations.

Virtually all science-based or, more often, as we have seen, technology-based societal issues do not entail one's knowing much about science or technology to recognize that a given technology or some application of technology may be detrimental to some segment of society; nor is it necessary to understand possible solutions in a societal sense, meaning in the sense of taking actions based on the advice of credible experts. We have already made the point that no *reasonable* amount of education in science can be expected to prepare the average individual to reach reliable, wholly independent judgments on societal issues that are strongly dependent on science or technology. Society must learn how to select and rely on the opinions of others who are more expert on such matters than are most of its members. More will be said on this point in a subsequent chapter, for it not only affords the only practical solution to the problem of scientific illiteracy, but also provides a firmer foundation for the STS movement.

We have also pointed out that while STS does not solve the problem of what science to teach to the general student, it does provide a basis for relating science to societal concerns, however remote the relationship may be in a given case, thereby offering students the freedom, which they do not normally have in science courses, of expressing their personal opinions. This opportunity for self-expression is an important aspect of STS that cannot be discounted. For many students science seems too prescriptive, offering no room for debate or personal views. The nature of science pretty much dictates such rigidity. We noted that all major disagreements and conflicts surrounding new *scientific* ideas have been pretty much resolved thorough detailed analysis *within* the scientific community before reaching the textbooks and classrooms of the general student. Modifications to existing theory and scientific laws continue, of course, but these are generally not at the level that can profitably engage such students.

Consider, as an interesting example, the modifications to Newtonian mechanics required by Einstein's special relativity theory and imagine Einstein's discovery as a contemporary event highlighted on the front pages of the local newspapers (as was done at the time in some newspapers), thereby becoming an obvious subject for discussion in the science classroom. What might be the nature of such discussion? Asking students whether they thought Einstein was right or wrong would be fruitless. The only possible responses would obviously be more intuitive than scientific and determined mainly by the students' experiences with nature, which would hardly include speeds at which relativity effects become noticeable. Thus, Newton would win the race hands down in the minds of students, but they would be wrong. Einstein's conceptual idea, as seen by the scientific community, and verified time and again through experiments designed to test its implications, turned out to be correct, despite its disagreement with the "commonsense" views held by most students. And so it is with most modern revolutionary ideas in science, for science long ago moved out of the realm of challenge by the untrained individual. Science cannot be decided by classroom consensus or by any democratic process; there comes a point where the general student simply must accept the collective opinion of the scientific community. While one might like to help students experience for themselves the discovery process by having them perform experiments designed to elucidate scientific laws, they all know that in the end the answers are already in hand, which serves to dull so many student laboratory experiences. Science for the general student is much too brief an exposure to permit detailed proof of everything that is taught, even assuming that the student were truly interested in such analysis. Hence general education in science, more than in most fields, involves a great deal of implied "trust me, this is the way it is," as teachers and textbooks are forced by the limitations of time and student interest to state conclusions without proof. This is where STS has a genuine advantage over conventional science programs; it provides a forum for personal views, not on science, to be sure, but on how it causes societal concerns, and gives the student a sense of involvement as a participant rather than merely an observer.

▲ MEASURING "SCIENTIFIC LITERACY"

We know that scientific literacy boils down to a matter of definition, but definitions also set standards for assessment. Once a curriculum reform movement sets its objectives, the obvious next question, of course, is, "How will you know when you get there?" As we have seen, in common with most educational experiments, it will take at least two decades to know whether the current wave of curriculum reform, under the broad heading of scientific literacy, has achieved any real success, and then only if a given program has some objective evaluation instruments built into it, or has clear enough objectives so that *meaningful* test instruments are readily designed. One way of measuring success, of course, is through objective longitudinal testing of students at different stages in their educational careers through high school and college. But this poses the problem of establishing a baseline now, which, in the absence of an agreed-upon operational definition of scientific literacy, is clearly impractical. Perhaps the current curriculum activity will spur efforts to reach a consensus on the meaning of scientific literacy, but there is no evidence that this is about to occur. Without a clear definition we must continue to resort to subjective and anecdotal impressions of the worth of a new curriculum. Not that these have no value; after all, it was largely because the science education community *felt* that the post-Sputnik reforms had failed to achieve certain perceived goals that the "crisis" was declared in the early 1980s;

there was no clear objective evidence of it. Thus there must be some vaguely defined norm that scientists and science educators subjectively attach to the meaning of scientific literacy and which they use to decide whether students have achieved this "elevated" mental state.

Reaching agreement on what shall be meant by scientific literacy is only part of the problem. Equally important is how it should be measured. Reliable evaluations of programs and meaningful assessments of student understanding are the most difficult stages of any new science curriculum development. Designing a new program is often easier than evaluating it properly. Selecting representative test sites, teachers, test groups, controls, evaluation designs, test instruments, and, in the end, interpreting the results unequivocally—all these require the input of experts. Not that there is a shortage of literature on the subject, for the field has been researched extensively. But this does not make it any easier for the novice, and too many educational experiments have suffered from poorly designed evaluations.

Testing for factual knowledge is simple and straightforward, and therefore is all that is done in most instances. But this is not the answer, for we know that facts alone, while necessary, do not constitute the essence of science and are certainly no measure of literacy in the subject. Moreover, failing continual use, most facts are soon forgotten. This prompts the question of what sort of *lasting* impressions of science can be conveyed to students, or, put another way, how can we achieve adult literacy in science? It is sometimes said that recognizing those who are *illiterate* in science is generally easier than identifying those who are literate, which is probably true of any specialized field involving cognitive skills.

This would suggest that any competent scientist or science educator, following a brief conversation with an individual, should be able to recognize whether that person is either totally illiterate in science or, as is more likely, knows something about science. The question then is to determine just how extensive this knowledge is. Granted that if the conversation is improperly directed by the science specialist, such an assessment could turn out to be overly subjective and misleading, just as a written examination or quiz may unfairly test an individual if the specific subject matter is outside that individual's range of immediate recall. But it needn't be so. Instead of opening such a dialogue with a general question such as, "What do you know about specific subjects in science?" the proper opening should be, "What is it about science that you would feel comfortable discussing?" This puts the student at ease, particularly if he or she comes prepared for such an oral quiz, and it is then possible to determine whether that individual is totally illiterate or possesses some level of scientific understanding. Moreover, one can also determine in this way whether such understanding is purely factual or includes a grasp of the nature of science, and particularly of the significance of its major conceptual schemes.

The conclusions reached by an interviewer through such a dialogue are somewhat subjective, of course, particularly if one seeks to obtain a quantitative assessment of the individual's level of literacy. However, one can easily determine in this way whether the individual is functionally literate in science, as this term has been defined in chapter 4, for example, or by any other agreed-upon set of criteria. Clearly, it is impractical to evaluate all individuals in such a one-on-one manner, but if one wished to sample a given population for scientific literacy, it might be far better done in this manner than through the usual "objective" test instruments. This is not to imply that written evaluation tools are of no value in science education, for they are certainly much easier to administer, and if properly designed they can elicit useful insights on certain aspects of scientific literacy. I am only saying that the immediate give-and-take of a personal dialogue can be much more revealing and is probably more equitable, especially when one is testing for the individual's general knowledge of what science is *about*, rather than for specific course content.

▲ DO EDUCATIONAL EXPERIMENTS EVER FAIL?

A basic problem with major educational experiments is that they take place over time, in most cases over a long time—years or decades—so the variables are difficult to control. It is one thing to test (objectively) on a contemporaneous basis whether a given topic in science is better presented in one way than another to properly matched cohorts, or by one textbook than another, or by one teacher than another, or by the same teacher to different classes, and so on. Because such experiments can be completed in just a few weeks or perhaps as much as a year, "aging effects" do not usually cloud the picture; that is, the teacher does not grow much more sophisticated in science over the duration of the experiment, the test sample does not grow up much, and the social and educational environments do not change.

But consider a new K–6 curriculum, or an entire high school curriculum rather than an individual course. Such extensive programs cannot be fully evaluated except over a period of years or even decades.

Then, trying to account for aging effects, as well as controlling the host of other factors affecting such experiments—differences among teachers, teaching methods, schools, facilities, sample validity, social changes—makes for an almost insurmountable task. And think of the problem of doing all this in a true longitudinal study, that is, by tracking *individual* students from the beginning of high school well into adulthood, which some believe is the ultimate test of a major new science curriculum for the general student. What is the likelihood in such a study of being able to attribute the end results to a single independent variable, namely, the curriculum? Most educators would regard this an impossible task, unless the outcome were so dramatic as to eliminate any doubt as to the cause. Small wonder that so many innovative programs are judged more on the strength of anecdotal testimony and personal impressions than on what most educators would consider convincing experimental evidence.

Educational experiments rarely fail, at least in the eyes of their proponents; but too many are borderline. The few that show such unambiguous results as to make it unnecessary to resort to statistical tests of reliability are clearly successful. These are "blockbuster" experiments. On the other hand, those that are so marginal as to require the assignment of confidence levels to their conclusions should be highly suspect, especially if the conclusions are called upon to support major changes in curriculum structure. What is the point in expending huge sums and endless effort on promoting major changes in educational programs if the statistical evidence for the worth of a given change is only at the few sigma level? Granted that one cannot predict in advance whether a given educational reform will have a blockbuster or a negligible effect, but through properly designed pilot studies, including *critical* evaluations, one might at least determine the *likelihood* of achieving a highly positive result. In the absence of such a possibility, it is doubtful that major curriculum reforms are warranted. Science experiments can also be marginal, which sometimes means that the experiments are improperly designed, or that the sought-after effects are indeed so small as to require extensive testing to obtain *reliable* results. In either event, a cost-benefit analysis based on preliminary findings informs the experimenter whether it is worth continuing. A great deal of time and money could be saved if educational innovators were to observe this practice diligently, rather than insist on carrying out their studies to the very end. One hopes that this sort of discipline will become the accepted norm in science education experiments.

The converse is equally valid; history should not be ignored in the search for new and better ways of presenting science to the public. Consider the new elementary science curricula of the 1960s and 1970s, discussed in the previous chapter. On the strength of both anecdotal reports and the statistical evaluations performed, these obviously were not blockbusters, but they could hardly be considered failures. Many teachers and science educators genuinely liked them, and the programs occupied center stage for a decade or two before gradually declining in popularity, for reasons having less to do with their merits than with the penchant of educational funding agencies, particularly the National Science Foundation, for blazing new trails. Today, the four major programs, SCIS, SAPA, ESS, and COPES, or variants thereof, can still be found in a number of school districts, particularly those that have strong science supervisors and a commitment to quality in elementary science education, as well as in some foreign countries. But regrettably, no more than about 5 percent of the U.S. elementary school population currently benefits from exposure to these programs, with SCIS accounting for the major portion. All the programs produced changes in student and teacher understanding. These changes were evidently not enough in the minds of those who control the educational purse strings to warrant their continued use in the face of newer programs that have different conceptual bases and less teacher dependence.

Why weren't the earlier programs revised to retain their best features, of which there were clearly many? And why weren't those that didn't seem to work as well replaced, as some individual school districts did on their own? Fortunately, SCIS has undergone a major revision through private rather than government funding, proving that there are at least some science educators and a commercial publisher who believe in the long-term potential of this well-designed program. The notion that a science curriculum can grow and improve with experience over time seems not to move curriculum developers and funding agencies as much as discarding the old in favor of new curriculum designs, most of which will eventually suffer the same fate. To be justifiable, in light of our experience to date, a newly proposed curriculum should show strong promise of provoking *major* changes in direction, not merely the incremental changes that characterize so many of the current curriculum projects. Mostly we seem to be getting new words to the same old tune instead of a brand new melody. Education must progress, of course, but after all the reforms of the '60s and '70s it is hard to imagine that *everything* now being done in the name of curriculum revision is really new or necessary. One wonders, too, how many babies are being thrown out with the bathwater.

▲ THE TEACHER REMAINS THE KEY

We have alluded several times to the unlikelihood of achieving a *marked* improvement in the scientific literacy level of the general public on the strength of curriculum changes alone. Even the best-conceived curriculum placed in the hands of a poorly trained or uninspired teacher is bound to fail. The problem of scientific literacy, or science awareness, or any other way one chooses to characterize one's state of mind relating to the scientific enterprise, while usually directly attributed to our schools, stems from a much larger societal problem. The teacher remains the key to quality and effective science education for the general student; yet on the whole, U.S. society continues to ignore this crucial factor.

All the finger pointing and school bashing cannot hide the fact that precollege education in the United States is not as highly valued as college education in the public mind, nor is it accorded the same professional status, which may result from the fact that being compulsory marks it as pedestrian rather than prestigious. We devote some 3 percent of our GNP to public precollege education, while spending twice that on defense and more than four times the amount on health care. This result may be taken either as a measure of our relative priorities, or of the greater strength of the health-care and defense industry lobbies. School budgets are routinely defeated in many communities, and the per student expenditures vary by a factor of almost three around the nation. Both of these facts reflect the wide range of interest in education, or lack therein, displayed by the residents of different communities.

The result is that many of our schools lack the total commitment and financial support needed to provide the highest-quality education to all students in every discipline—and especially in science. It is not simply a matter of teachers' compensation, although this is one manifestation of the larger problem, but rather one of professionalism. Elementary and secondary school teachers often complain of having little voice in the selection of curricula, textbooks, and other academic decisions, and of occasional interference from parents and school boards. School principals control virtually everything that goes on in their schools, they complain, and in some cases choose not to consult the teachers in decisions that affect their professional well-being. While in most school districts teachers are encouraged to continue their education, a major incentive being automatic salary increments, they are not often afforded the opportunity (time or travel funds) during the school week to attend professional meetings or otherwise keep up with current developments in their fields. Hence only the most conscientious and dedicated science teachers manage to bootstrap themselves to a level of competence that may truly be considered professional.

The other side of the coin is that professional status is rarely placed high up on the list of issues in collective bargaining negotiations, giving way to the more pragmatic concerns of salary, fringe benefits, and general working conditions. Suggestions that a system of ranking teachers according to competence be established, somewhat similar to that practiced at the college level, are invariably rebuffed by teachers' unions, fearing that any system of advancement that depends on peer evaluation of teachers, regardless of who the evaluators may be, has the potential for nepotism and politicization. They prefer automatic rewards, based on in-service and/or continuing-education credits, which afford less opportunity for favoritism but also do little to promote professionalism among teachers or to recognize that there are differences in the quality of tenured educators.

Ernest Boyer said in his report on secondary education in America a decade ago, "Two of the most troublesome aspects of the teaching profession are the lack of a career ladder and the leveling off of salaries."[5] Since then, teachers' salaries have improved markedly in most school districts, with the result that many communities now complain about excessive teacher salaries and vent their discontent by rejecting school budgets. Effective career ladders and professional recognition still have not improved much, which in the long run would probably have benefited teachers (and students) more than any of the short-term gains. The fault does not lie solely with the teachers or their bargaining agents, or even with the communities. Much of it can be attributed to our universities, and particularly to the teacher training institutions, which have failed to win for precollege teaching the recognition it deserves. High school teaching as a career goal is often disparaged by the faculties of our liberal arts colleges, who tend to look down upon schools of education as being intellectually barren, the end result being that the most able students generally do not seek teaching careers.

Compounding this problem is the troubling fact that those who determine the education of prospective teachers, namely, schools of education, also control their subsequent credentialing by state education departments, a system obviously lacking the basic checks and balances needed for serious upgrading of the profession. Certified does not necessarily mean *qualified,* and it is well known that teacher certification requirements vary widely among the states, a situation that is particularly noticeable in science and mathematics. Several years ago, in an effort to create a single, independent set of standards for science teachers, the National Science Teachers Association (NSTA) established a voluntary national certification program to recognize those teachers who met its standards. They hoped this would lead to an enhanced sense of professionalism, both in the self-esteem of the individual teacher and in the eyes of local and state education authorities.[6] Unfortunately, the program has attracted relatively few applicants to date, probably because, as it is unofficial, most teachers fail to see enough personal benefit in it, and government bureaucracy being what it is, state education departments are not inclined to accept NSTA certification in lieu of their own. One can quarrel with the NSTA program on the same ground on which state certification programs are often faulted, namely, that the procedures are not very rigorous because certification is usually based solely on the teacher's academic record and experience, without formal examinations or other direct means of establishing the applicant's actual knowledge and ability. This will undoubtedly change in the next few years, however, as the states move toward more stringent requirements, including written examinations in one's discipline, particularly in the sciences.

What all this comes down to is that neither the universities nor the teachers' professional organizations have managed to convey the message to the community at large that teaching is a serious business and that teachers must be treated as full professionals if we wish to see a marked improvement in the quality of education in America, certainly in the specialized field of science education. These days one often hears the *serious* suggestion, no longer in jest as it was at midcentury, that the best way to reform the education of prospective teachers and encourage liberal arts faculties to take a greater interest in pre-college education generally would be to abolish all undergraduate schools of education and establish in their place departments of education in liberal arts colleges. Indeed, in a memoir written after he was forced to resign, Terrel Bell, secretary of education in the first Reagan cabinet, who rebelled at his assigned mission of abolishing the department, recommended that no undergraduate degrees be awarded in education, only postgraduate ones.[7] There is obviously some merit in the idea, at least for prospective high school teachers, who must in any event complete at least the equivalent of a major in their chosen discipline if they are to become effective teachers. If widely adopted, this move could go far toward enhancing the teaching profession. Some universities already have such a system, and barring a truly *substantive* reform of schools of education with respect to the training of teachers, it is likely that the practice will spread to many others, which would be a boon for science education.

There must be an axiom of education that at some point in a child's development, teachers cannot effectively teach what they themselves do not fully understand. Where this point may be in science education is not totally clear, but certainly at some grade level in the elementary school the effective teacher *must* understand more about science than can be conveyed solely through teacher's guides and a limited number of in-service institutes. This poses a host of questions about science in the elementary schools—questions regarding the preservice and in-service science education of elementary school teachers, of whether, in fact, it is at all possible to achieve a reasonable level of science sophistication in the more than one million elementary school teachers in the U.S.; or whether we will ultimately need science specialists in all our elementary schools. All of this bears directly, of course, on the basic question of whether scientific literacy really begins in the classroom or in teacher training institutions.

No matter how it may be defined, scientific literacy as an *immediate* goal of elementary science is obviously an unrealistic expectation, but not so much for want of good curricula. Almost any activity-centered inquiry program in the hands of a "qualified" science teacher, meaning one who is reasonably literate in science, that engages students for more than the bare minimum of time normally allotted to elementary science each week in the U.S. should produce impressive results. As we already know, however, very few elementary school teachers, through no fault of their own, are truly qualified to teach science. Their preservice education in science, when compared with the many other subjects they must deal with in the elementary grades, is wholly inadequate to prepare them for the task. In fact, most elementary school teachers in the United States have had no science beyond their high school days. And this science, in many cases, consists only of a course in biology—and possibly one in general science. Even with the in-service courses or summer institutes in which some teachers manage to enroll, which undoubtedly add to their self-confidence by putting them in direct contact with science professionals even if adding only in a limited way to their understanding of the nature of science, there is no escaping the conclusion that few elementary school teachers are truly prepared to teach science.

Hence, if the elementary school is to be a fertile ground for sowing the seeds of some form of scientific literacy, as many science educators, this one included, believe it could be, we have three options. Either we continue to design elementary science programs to be as teacher-proof as possible (which appears to be a guiding principle of most of the current programs), in which case we should reevaluate our goals for science at this level; or we restructure the preservice education of elementary school teachers; or we decide that science education is important enough to justify the expense of having trained science specialists in all of our elementary schools.

Not too long ago, in the more affluent days of elementary education, it was not unusual to find science specialists in some elementary schools, particularly those in which the principal or the community understood that good science teaching requires special training or aptitude on the part of teachers, just as does art or music or physical education. In such schools all students were taught by the science specialist, either by rotating them through a dedicated classroom or by rotating the specialist through all the classrooms for specific periods and encouraging the regular classroom teacher to do follow-up activities. This worked well. But as school budgets were trimmed in the last two or three decades, elementary science specialists in the public schools were among the first to be considered expendable. Yet if a science specialist were added to *every* elementary school (K–6) in the nation, the per pupil cost of education at this level would increase by no more than about 2 percent. Putting this amount into a currently meaningful context, the total annual cost would be about equal to what we spend in one day on health care in the U.S. (roughly two billion dollars). This could be the most effective single step we could take to alter dramatically the pattern of elementary school science. Training the 50,000 or more teachers needed for this role would be no small undertaking, of course, requiring many years and a restructuring of teacher training institutions, but if the nation is truly serious about science education it should do no less. Except for the science specialists, the other "solutions" suggested here are palliatives, not remedies. The only true remedy, for which there is no viable alternative, involves better-trained teachers.

There is another, less costly way of achieving a similar end, which is to recruit high school science teachers or highly capable science-bound seniors as elementary school science "mentors." Their specific role would be to assist the elementary school teacher during the class periods devoted to science, preferably in the elementary school environment, but possibly in the high school itself. This has been tried on an ad hoc basis in some school districts with mixed results. Obviously, it presents many problems of an administrative nature, as well as some financial implications, either in terms of additional payment or released time for the mentors, where they are high school teachers, and course credit or the equivalent in community service credits for the seniors. A further benefit of enlisting capable high school seniors in such a program is the possibility that the experience may lead some into teaching careers. Using high school students as mentors also has the advantage that elementary school children are often more comfortable working with other young people than with adults, particularly with high school teachers. While the latter have a better grasp of science than the elementary teachers, they may not relate as well to very young children.

Probably the main reason why this approach has not been tried on a large scale is the lack of a strong organizational commitment, including funding, at least districtwide if not statewide. It would be a great service if some responsible educational group would undertake the development of guidelines covering such a program, so that a meaningful demonstration could be carried out. Until this teacher problem is somehow resolved, elementary school science will continue to suffer more than any other segment of our educational system, for at this stage of curriculum development it must be acknowledged that the teacher, not the curriculum, is really the key. Adding to the importance of such moves is the uniform observation that at the elementary level practically all children quickly become enthusiastic about science; it would be unconscionable not to build on this interest before it is lost in the high schools.

▲ REPACKAGING SCIENCE EDUCATION

Curriculum reform and better teacher training may not be the only solutions. The more pragmatic educators recognize that curriculum reform alone, without a better understanding of how students are motivated and how they learn, leads to a dead end in science education. Here the answer to some educators lies not so much in the science curriculum itself but on how it is packaged and delivered. The political scientist Sheila Tobias believes she has found some clues to why so many students are turned off by science. Using (nonscience) college professors and graduate students as her test subjects, such individuals being considered well motivated and intellectually capable, Tobias concluded that much can be learned on how to package science education by listening to the "students," to their complaints and reactions regarding conventional science courses, and to their suggestions on how the course structure might be improved.[8] Her research is unique and noteworthy, but anyone hoping to find the magic formula in these studies will be disappointed. Beyond the obvious question of how far one can extrapolate from such a specialized test sample, it is clear from the subjects' responses that many of the problems they perceived with science actually relate to the nature of the discipline, to its vertical structure, to the role played by mathematics, to the extensive use of conceptual models—in short, to the very characteristics that we have already described as making science seem hard to the uncommitted student. Tobias seems to have rediscovered the problem, but regrettably not the solution. This is not to say that we cannot learn from well-designed studies aimed at real-life students, but if the answer is merely to simplify the *way* that conventional science is presented we will not have made much progress.

Repackaging science education to make it more attractive or acceptable to students has been tried many times, always with results that in the end leave thoughtful educators wondering what sort of lasting impression of science, if any, the students actually derive from such courses. The teacher-student interaction is an important part of any educational experience, and while there is undoubtedly some truth in the complaint one sometimes hears that science teachers tend to be more dogmatic, more impatient, and even more arrogant than other teachers,[9] this is hardly the answer. All academic disciplines, the humanities as well as the sciences, have their elitists who simply tolerate students rather than engage them in a rewarding intellectual adventure. Fortunately, most teachers do not fit this bill, not even in science, where because of the peculiar nature of the discipline there may be more reason to find self-confidence, and even hubristic behavior, among its practitioners.

In a subsequent study, Tobias examined several undergraduate curriculum reforms, seeking common elements to account for successful programs.[10] She makes the important, though obvious, point that for an effective science program, dedication and commitment on the part of both faculty and administration (particularly at the departmental level) are essential. She adds, not surprisingly, that successful undergraduate science programs are most likely to be found in four-year colleges, where the science faculties, having no graduate students to deal with, tend to form closer mentor relationships with their undergraduate students. Unfortunately, her assessments of various ongoing programs suffer somewhat from the fact that they are based mainly on contemporaneous anecdotal evidence rather than on more objective longitudinal data. Nevertheless, her conclusions accord with the subjective impressions of many undergraduate science educators and seem entirely credible.

Tobias focused her studies on the undergraduate college level, where innovation and change are easily accomplished in course and curriculum design, certainly for nonscience students and even, though to a lesser degree, for science majors. It is interesting to note that the one other area where comparable freedom exists for the practice of new curricular activities is the elementary school, where there are also few vested interests or controls standing in the way of educational experimentation. The situation is quite different at the high school level, where state departments, often guided by the advice of local universities, rigidly control the high school science curriculum, at least those courses that may be regarded as prerequisite to comparable college courses, or that qualify the student for advanced placement in science. More freedom may be permitted for courses intended for the general student, but even here one is generally limited by the state's science requirements for high school graduation. Hence one should not expect to find *marked* changes in the high school science curriculum, but mostly the same books in new covers. While this may be acceptable for science-bound students, the nonscience students, who make up roughly 90 percent of the high school population, are not well served by it.

▲ THE SCIENCE-BOUND STUDENT

Our concentration on scientific literacy for the masses sometimes blinds us to the fact that we must also be concerned about the education of a small group of students, the 5 percent who are our future scientists and engineers. Are they being short-changed in this process? Somehow, in our rush to solve the mass education problem, we have assumed that all is well with the science-bound student, and that to solve any scientific personnel shortages, if they occur, we need only get more science-bound students into the pipeline, or keep those already committed to science from dropping out. But focusing on quantity rather than quality may turn out to be a serious mistake.

Judged solely on the overall quality of the end result, relative to the international community, American scientists and engineers rank at the very top in almost all fields. This is certainly true of the natural sciences, where the U.S. has dominated the Nobel Prize lists for the past half century. It is also true in engineering, where, while Japan has made great strides in the postwar period, particularly in the electronic and automotive engineering fields, the U.S. must also be counted a world leader, owing principally to the quality of its college and postgraduate education. The question of quantity, of whether this nation is attracting enough students into science and engineering, is quite another matter, one that I discussed in great detail in the opening chapter and concluded that the picture is not as grim as so many manpower studies and commission reports have made it out to be.

Nevertheless, there is cause for concern. The science and engineering professions have declined in popular esteem in the past two decades, perhaps partly the result of premature publicity given to some scientific claims that later turned out to be false (e.g., most recently the "cold fusion" fiasco) and revelations of instances of actual laboratory fraud or misconduct in science, all, because of their relative rarity, being headlined in the popular press. Not that the scientists involved were blameless, but as a result the entire enterprise has come under a microscope, making it more difficult to carry out its normal self-correcting procedures before politicians and the press have had their field day and public opinion has been negatively influenced. It is doubtful that public perceptions turn many committed students away from science, but they must have some impact on the uncommitted and, as everyone knows, science does need friends in the community at large. How to win such friends—indeed, whether they can be won—will be discussed later. However, we should not assume that because the number of nonscience students is so large compared to the science-bound, or our professional scientists and engineers are so highly qualified, we can be complacent about this committed segment of the student population.

It would be mindless to believe that we have reached the ultimate in quality education for the science-bound student, that we cannot hope to produce even better-qualified scientists and engineers in the future, or that we cannot make the science programs, facilities, and teachers in the schools and colleges enough of an attraction to ensure a steady supply (without special inducements) of very high quality science and engineering professionals in the future. From the relative numbers of students involved, one can estimate that perhaps 85 percent of the available science teacher time and facilities in our high schools is devoted to teaching the general student and the balance to the truly science-bound. At the college level this percentage is much smaller, at least for senior faculty, but the fact remains that a good deal of time and energy are given over to the nonscience student.

An obvious, though seemingly disparaging, question might be: "To what end?" If general education in science fails to produce literate adults, why spend so much time and money on it? Might it not be wiser to devote a greater share to educating the science-bound, where it could be more cost effective? The question is purely hypothetical, of course. Educational politics aside, we have already alluded to the many problems such a move would cause, not the least being an immediate surplus of science educators and teachers and the wrath of the science education community. But the question does prompt us to consider seriously the purpose of science education. I have previously pointed out that the most committed students somehow survive the worst in our educational system: poor teachers, poor facilities, and, in some instances, even lack of family support; but they nevertheless manage to become scientists and engineers. However, not all science-bound students, despite their interest in the field, are this strongly committed, and these can be turned off by poor teachers or inadequate facilities. If we are truly determined to produce the best possible scientists and engineers, such students deserve every possible advantage, including access to the best teachers and to specialized science high schools at the state or even national level, when suitable facilities are not available locally.

Do we want more students entering the science/engineering pipeline? Then make the educational opportunities in these fields more attractive to them (and their parents) at the high school level. Do we want to keep them in the pipeline through college? Then make the professional incentives and career opportunities so attractive as to dissuade them from defecting. These are easier said than done, of course, but should the problem of attracting and retaining students in the science/engineering pipeline ever turn critical, such measures would quickly be put into effect.

▲   THE FORGOTTEN GENERATION—
INFORMAL EDUCATION

Even assuming that some of the precollege curriculum reform efforts now underway succeed in capturing the interest of the general student, would the problem of scientific literacy be solved? Obviously not, for unless this momentum carries through the college years and into productive adulthood, we cannot truly say that the main objective of science education for all citizens has been achieved. The point has been made several times that if science is to penetrate the consciousness of our society, what we must seek is a literate adult population, not merely literate students. Hence the problem is not only one of curriculum reform at the precollege level; we must also be concerned about what happens in the students' college years and beyond. If a curricular approach that shows promise in high school is not reinforced in college, whatever gains may have been made for the general student in high school science are likely to be lost.

Clearly, for any scientific literacy program to succeed it must encompass the full educational spectrum, including the adult years. This is not an easy task, for where precollege education is largely coordinated, at least on the state level, college education is not, each institution having its own distribution requirements. It is not easy to get college science departments to agree on distribution requirements for the general student, and more difficult still to persuade a full college faculty to radically alter its overall course requirements. A major deterrent in this regard is the widely held view that college students ought to be free, within reasonable limits, to choose their own courses. This, coupled with the well-known custom of science departments to draw in students through attractive course titles, or by announcing no mathematics requirement, or by gaining the reputation for "easy" courses, makes it very difficult for a faculty to agree on a *single* science curriculum for all students. Hence it is entirely possible that a promising new high school approach to science education can be defeated in the end for want of general acceptance at the college level.

There is the further problem of how or whether to develop an effective system of informal science education, both to reinforce student awareness of science in adulthood, and to develop such awareness in those adults who were exposed only to traditional science during their school years, or were turned off by their experiences— those who might be termed the "forgotten generation." Several avenues of informal education exist for this group, but none has yet proved effective. Early in its existence, the NSF recognized that to carry out its mission it must support efforts to achieve a better public understanding of science, and it established a special office to address this goal. Many millions were subsequently allocated to fund a variety of "public understanding" programs (now called "informal education") with little to show for it in the way of positive results.

George Tressel, formerly director of NSF's Division of Materials Development, Research and Informal Science Education, now concludes that most of the public doesn't want to know about science: perhaps 20 percent are "attentive" to science (meaning that they *might* be attracted to TV programs like *Nova*, if such programs were properly "packaged"), another 20 percent find science interesting but do not pay much attention to it, and the remaining 60 percent have no interest at all.[11] Tressel believes further that the best way to increase the attentive public is to get to children at an early age. If Tressel is correct, and there is no denying that young children are very easily attracted to science, it would mean that there is little point in trying to reach adults directly through informal education means. However, there is no convincing evidence that bringing science to children results in their becoming more attentive adults. Despite Tressel's contention that *everything* has been tried in the way of informal education, without much success, one wonders whether there may still be productive areas that have not been thoroughly explored.

Getting the general adult population more attentive to science—of trying to double, for instance, the 20 percent that Tressel believes presently fall into this category—is a massive undertaking. One thinks immediately of the mass media, both print and video, as being the ideal means of communicating science and technology to the public, and so they might be if editors, publishers, and TV executives believed that their responsibility included public access to education as well as to news and entertainment, which for the most part they do not. Or, if the business/industry community believed in supporting such educational efforts as a public service form of advertising in the mass media, which it does only to a very limited extent, the pattern of informal education might undergo a profound change. But lacking a broadly endorsed national drive for science awareness (or literacy), neither of these is likely to occur.

The occasional TV series or newspaper article do not provide the sustained exposure to science that would be needed to gain the public's attention. Regular science (not health) columns, once fairly common, have disappeared from most newspapers. In fact, it is probably fair to say that fast-breaking science news stories aside, science on a regular basis gets less total newspaper space in the U.S. than does astrology. Moreover, it does not appear likely that this bizarre pattern will change in the foreseeable future. Yet, if awareness and appreciation of science by adults is to be our goal, as is suggested in the next chapter, a possible route may be through *informal* education, particularly by the mass media, which have a real opportunity to effect a change in the way that society looks upon science. It also holds the key to reinforcement of science facts and concepts for educated adults in the future. Unfortunately, the popular press for the most part gives space only to the "newsworthy" in science—the spectacular, the amazing, the commercial aspects of science—not to the task of simply helping educate the public. If newspapers gave as much space to science columns on a regular basis as they devote to the pseudoscience of astrology, it is conceivable that our society would not be quite as illiterate in science as it is today. The National Science Foundation has in the past encouraged partnerships between educators and book publishers to develop specialized science text materials for the elementary grades. It might do well to consider forming similar partnerships with a few major newspaper chains and large-circulation magazines as an experiment in informal science education.

▲ ADULT EDUCATION IN SCIENCE

What are the prospects of a widespread movement toward evening adult education courses in our high schools and colleges? Continuing education for adults has long been available in classroom settings and normally attracts a small but dedicated group of individuals, many of them retirees and only a few interested in science. While it is less formal than a degree program, there is still something about classroom-based education that fails to attract large numbers of adults to science courses. The incentive even for this relatively painless form of science education is obviously lacking.

Public libraries have been proposed as sites for informal education in science, not in the way of formal courses or lectures but more as resource and information centers having special areas ("science corners") stocked with print and video materials relating to science and technology, and with special librarians experienced in dealing with such materials (who might be high school or college teachers moonlighting after school hours). The advantage of such informal learning sites is the easy availability of public libraries around the country, all of which have some sort of science collection and many of which have on-line facilities to access other, larger collections or can arrange to obtain special material by mail. A disadvantage, of course, is the cost of expanding existing facilities, including hiring specialized librarians. Unfortunately, libraries and schools seem to be among the first to suffer the effect of tight budgets in most communities. Communities should also take note of the number of retired scientists and engineers who, in the proper setting, would be happy to bring their knowledge and experience to bear on the informal education of adults—at minimal expense. This would seem to be an ideal opportunity for local business/industry groups to underwrite experiments in adult science awareness at a modest cost. Such experiments, if properly monitored, could provide useful information on just how attentive to science we might be able to get our adult population, if it is offered to them in such an easy, unpressured manner.

Probably the most logical sites of adult literacy or awareness programs are the increasing number of science museums, generally aimed at children, that have been springing up around the country. Most offer the advantage of actual hands-on experience with science and technology, of the kind that clearly captures the interest of children and appears to be equally appealing to many adults. Some, in fact, offer workshops for adults; others have multigenerational programs for adults together with children. But the major problem here is the scarcity of such sites: only about 150 museums in this category presently exist in the U.S., and they are costly both to establish and to maintain. Another problem, relating to both museums and libraries, is the challenge of getting some reasonable fraction of the adult population interested enough in science to devote time to it at either place, and this would require a major public relations campaign. Thus, the only conclusion one can reach from all this is that while there may be some promising approaches, there is no easy way of directly tackling the issue of adult literacy in science, short of an all-out drive aimed at this target. But even under the best conditions it is unreasonable to expect that large numbers of adults would respond to such a drive.

The question, then, is how "cost-effective" such an effort might be; how many (actively) literate adults would make a significant difference in the way the scientific enterprise is regarded by our society? Probably, no more than about 15 to 20 percent of the adult population, or roughly three to four times the present number of professional scientists and engineers (see chapter 8). These are not merely the "attentive" science patrons referred to by George Tressel, but much more; they must be fully aware of the *nature* of science and of its role in modern society. By any reasonable definition, such individuals would be classified as scientific literates.

With such a cadre of knowledgeable individuals scattered throughout the population, the chance of finding voices of reason in community debates having something to do with science or technology, including the role of science in its educational system, would be greatly enhanced and both sides might rest easier. On the one hand, the scientific community should be satisfied that its activities were not coming under unreasoned attack by unknowing or politically motivated groups, and on the other, the community should feel more comfortable with its actions for having such a resource within its ranks, sharing its concerns and helping guide it to rational decisions. It sounds easy enough, but of course it is not. Getting even this small fraction of the adult population to be sufficiently literate in science to the point of playing such a catalytic role is an enormous task, if at all possible. The chance of accomplishing this outside the normal school environment is very slim indeed, far less than with a properly designed formal education.

CHAPTER 8

How Much Scientific Literacy Is Really Needed?

Knowledge which is acquired under compulsion obtains no hold on the mind.
—Plato (*The Republic*, Book VII)

▲   We have seen that short of a paradigm leap in how society regards the importance of becoming (and remaining) knowledgeable in scientific matters, it is unrealistic to believe that there will be much change in the public's personal understanding of science in the foreseeable future. Some, of course, notably the postmodernists and counterculturalists of science, would assert that what society really needs is a drastically new theory of knowledge to take us out of the current mode of cloning scientists who have the same sets of values as their teachers, and the same ways of doing science. The latter, we know, is unlikely to occur as long as science continues to show success, for except to please some social reconstructionists there would be no good reason for it. That science must somehow find an accommodation with society—to the extent that sociologists speak for society at large—with modern social theory is a given, but not if this means a major restructuring of science. This, however, is not the real issue, which is getting the public to feel comfortable with science. The real question is whether we can gain more by redefining the *purpose* of public education in science than by continuing to nibble at the edges of science education.

▲   THE TWENTY PERCENT "SOLUTION"

We have seen that the massive reform movement currently under way to improve the nation's scientific literacy is flawed in several respects. Its chief flaw is its failure to (a) set agreed standards for such literacy,[1]

(b) establish convincing guidelines for achieving these standards, and (c) specify how such improvement, if actually attained, would benefit society any more than would enhanced literacy in any one of a number of other academic disciplines, such as political science, economics, psychology, or law, disciplines that more directly affect the ordinary affairs of a society. The lack of focus in this reform movement appears to be justified by the science education community mainly on the ground that any improvement, however small, will serve the overall good of students as well as of science educators.

Until very recently, two separate grounds were regularly cited in support of the reform movement in science education, the first being a concern that the nation was facing a serious shortfall of scientists and engineers (and science teachers), which could be offset only by improving the science education of all students. This alarm, we now know, had pretty much propped up the science education community for a decade, since it was first sounded so urgently in 1982.[2] Coupled with this was the argument that because of the many controversial societal issues having a scientific (usually meaning technological) base these days, the public must be made more literate in science to deal rationally with such matters. It has already been shown how both arguments fail the test of simple logic,[3] but in particular the former, based mainly on the National Science Foundation's "pipeline" studies, is now widely recognized as being faulty.[4]

Thus, with the "manpower" issue no longer dominant, the purpose of the scientific literacy movement takes a somewhat different turn. This leaves the societal issue as the main rationale for seeking widespread literacy in science, which may seem to simplify the problem but is actually a more formidable objective. Let us assume, therefore, that what we seek is scientific literacy in the *social* sense; that is, we would like people to understand (a) how the scientific enterprise works in our political and economic climate, and (b) how the public can reach independent judgments on science- or technology-based societal issues. The reader will be quick to recognize that this definition is deceptively simple, for while the first part may be readily achievable, the second takes us full circle to an understanding of science itself, which we have seen is beset by a great many problems. Nevertheless, this may be the only *practical* definition of the elusive scientific literacy, although how one measures it remains pretty much a mystery, as does the question of how science education might be restructured to achieve such literacy in an appreciable number of students. In any event, assuming this were the target a theoretical "what if" exercise suggests itself as setting some bounds on the problem.

How much scientific literacy is enough in a democratic society? Few responsible educators really believe that any amount of reform or tinkering with science education will ever elevate all, or even most Americans, to any reasonable state of scientific literacy, however one chooses to define it. We are nearing a century of experience in the U.S. with general education in science, both at the secondary and college levels, and have little to show for it by way of public literacy in science. Perhaps it is time to give up the idea that such literacy can be achieved merely by exposing all students to some form of compulsory science education, rigorous as that might be made, and hoping that enough of it sticks with them to make us a scientifically literate nation. If the past is any harbinger at all of the future, we have no reason to believe that this is feasible. However, it turns out that a more modest improvement in the scientific literacy rate could have a marked influence on public attitudes regarding science in our society and on the outcome of public debates.

Putting aside for the moment the question of how scientific literacy can best be defined in a meaningful (and measurable) fashion, we turn to the somewhat simpler question of how many Americans may be considered scientifically literate by *any* but the most trivial standards. Here we find estimates ranging from roughly 3 to 6 percent of the adult population. The lower number includes only trained scientists and engineers, on the simplistic notion that at least this group ought to be considered literate in science. We know, however, that many individuals trained in science or engineering lack the skill or interest in societal matters to be able to contribute constructively to technical controversies affecting the public. Hence the fraction based on this group is likely to be less than 3 percent. The higher estimate of 6 percent derives from Miller's benchmark studies.[5] Thus, while still possibly on the high side, it seems fair to conclude that perhaps as much as 5 percent of American adults may be sufficiently literate in science *and* social mores to reach independent, rational judgments on technology-based societal matters.

It is reasonable to expect that the presence in an interactive group of one or more scientific literates (who satisfy, for example, the above independent judgment criterion) should help to inform and even guide the overall group toward reasoned conclusions and actions on issues involving science or technology. Not in the guise of outside experts,

▲   **Table 8.1** Values of $P_{\geq 1} = [1 - (1 - f)^N]$ as a Function of Group Size (N) and Literacy Rates (f)

	$f = 0.05$	$f = 0.10$	$f = 0.20$
$N = 6$	0.26	0.47	0.74
$N = 12$	0.46	0.72	0.03
$N = 25$	0.72	0.93	0.996
$N = 50$	0.92	0.995	0.9999

who are often resented by members of a deliberative body, especially when they suspect that the expert was retained to espouse a particular point of view, but simply as fellow group members (colleagues, family, friends, neighbors, etc.) who happen to have some specialized knowledge bearing on a topic under discussion. Even if such individuals were not inclined to assert their knowledge or experience to other group members, at least they might be expected to focus the group on the real issues and deflect unfounded rumor and speculation. Think of the effect on how society deals with science-based issues if every group considering such matters had such a built-in "knowledge base."

The question then is, assuming that in the overall population a fraction f satisfies a particular literacy criterion, what is the probability $P_{\geq 1}$ that in a group of N individuals, one or more will be literate in science? Simple probability theory yields the expression:

$$P_{\geq 1} = [1 - (1 - f)^N].$$

Shown in table 8.1 are values of $P_{\geq 1}$ as a function of group size N, assuming the group is randomly selected from a normal population. The dependence on literacy rate f is shown for the values .05 (which we have assumed is near the actual value in the U.S. today), double this rate (.10), and doubled again (.20); that is, literacy rates of 5 percent, 10 percent, and 20 percent.

Except for groups approaching fifty or so members—such as a town hall or community meeting debating some environmental issue, or a tenants' meeting in an urban apartment dwelling—the probability of finding at least one scientific literate in the group is not very high for $f = 5$ percent. For instance, in a small family gathering of six members, the chance of finding one individual literate in science is only 26 percent, meaning that about one in four such family groups would be expected to include a member who is literate in science. Similarly, in a group of twelve members, as in a trial jury, there is only a 54 percent chance of finding one or more such individuals; that is, on average, roughly one in two jury panels would have such expertise. Even in a fifty-member group, while the probability is fairly high (92 percent), there is still no *certainty* of finding such a literate in the group. More will be said about the special case of juries later, but one can easily see that for a typical randomly selected committee or panel, or any fact-finding body numbering fewer than a dozen members, it is unlikely that anyone in the group will be competent to reach an objective *independent* judgment on a science- or technology-based issue, or to help fellow members form knowledgeable opinions on such matters. Thus, the conventional wisdom that to be manageable committees ought to be small may actually work to society's disadvantage when the committee's charge involves technical matters and the group is not specifically selected for expertise in the particular subject area being considered.

Doubling the literacy rate to 10 percent of the population improves the odds markedly, of course, but still not enough to take comfort from the belief that scientific literates will now be found in all reasonably sized groups. Our six-member group, for example, would still have only a 47 percent chance of including at least one scientifically literate member, and the twelve-member jury panel would have a 72 percent chance. Contrast these improved yet still rather slim expectations with the chance of finding one or more such literates in a fifty-member group, which now becomes 99.5 percent, virtually a certainty.

On the other hand, doubling the literacy rate once again—to 20 percent of the general population—makes it likely that even smaller groups, such as governmental committees, will include at least one scientifically literate person. On our twelve-member jury, for example, the probability now becomes .93 (a 93 percent chance). With this literacy rate we are almost certain to find at least one literate in groups as small as twenty, where the probability is .99. Even in our six-member family group or a civil jury, the chance that at least one will be literate in science is now 74 percent. To have a 99 percent chance of finding at least one scientific literate in a random group of twelve would require a national literacy rate of about 32 percent, and for a group of six, a literacy rate of 54 percent.

However, the requirement that only *one* or more individuals in a group be literate in science, which we have seen yields reasonable

▲ **Table 8.2** Values of $P_{\geq 2}$ as a Function of Group Size (N) and Literacy Rates (f)

	$f = 0.05$	$f = 0.10$	$f = 0.20$
$N = 6$	0.03	0.11	0.35
$N = 12$	0.12	0.34	0.73
$N = 25$	0.36	0.73	0.97
$N = 50$	0.72	0.97	0.999

expectation values only for large groups (about fifty members), given the current U.S. literacy rate, may not always satisfy society's needs. Group dynamics also plays an important role. Having only one participant in a large group who is literate in science by our criterion, but is not a recognized expert on the particular issue being considered, will not necessarily comfort those members of the group who have a fixed contrary opinion based on rumor or personal belief. Having two such literates in the group would be more convincing, of course, provided they agree with each other. Should they not agree, as often happens when assessing the societal impact of technology, rather than having the group decision turn on their relative rhetorical skills the preferable alternative would be a third such literate. The probabilities ($P_{\geq 2}$) of finding two or more in the group are shown in table 8.2 for the same group sizes and literacy rates. Consider a fifty-person assemblage, which we have seen has a 92 percent chance of including one scientific literate (at a 5 percent literacy rate). Table 8.2 shows us that such a group is far less likely to have more than one such literate in its midst.

Similarly, table 8.3 shows the probabilities of finding three or more literates among these groups. Note that while one is virtually certain of finding two or even three literates in a group of fifty individuals,

▲ **Table 8.3** Values of $P_{\geq 3}$ as a Function of Group Size (N) and Literacy Rates (f)

	$f = 0.05$	$f = 0.10$	$f = 0.20$
$N = 6$	0.002	0.016	0.10
$N = 12$	0.02	0.11	0.44
$N = 25$	0.12	0.46	0.90
$N = 50$	0.46	0.89	0.999

given a literacy rate of 20 percent, the probability remains poor of finding such representation in groups smaller than twelve. The chance of finding two literates in a group of twelve, for example, is 73 percent, while in a six-member group it is only 35 percent.

Some caveats obviously apply to the foregoing analysis, since it assumes that group members are *randomly* selected from the *average* population. In practice such an idealized model is rarely found. Geographical factors can skew the outcome appreciably. In an academic environment, for example, or in the neighborhood of a large research center, the scientific literacy rate in the local population would likely be greater than the average 5 percent we have assumed for the nation as a whole. Similarly, in some rural or inner city communities, where higher education may not be the norm, the scientific literacy rate would likely be lower than 5 percent. Also, the criterion of perfectly *random* selections may not always be satisfied since the makeup of citizens' groups often must satisfy some political, racial, ethnic, or economic considerations; hence they cannot be purely random.

Courtroom juries represent a special case. Here the selection is very nearly random, with potential jurors generally chosen by lot from voters' lists, until the very last stage, the *voir dire,* where one or another attorney will invariably seek to challenge potential jurors who they believe have independent knowledge of a technical area that may be at issue in the trial, or who they feel may not be readily convinced by their own "expert" witnesses. The net result is that scientific literates are almost always excluded from juries hearing cases in which their *independent* judgments might be brought to bear, a practice that often invites critics of the jury system to agree with Mark Twain's wry observation that "the jury system puts a ban upon intelligence and honesty, and a premium upon ignorance, stupidity and perjury."[6]

The foregoing analysis shows that if our national scientific literacy rate, defined here in the pragmatic sense of individuals able to cope independently (and rationally) with science-based societal issues, were in the order of 20 percent rather than the 5 percent (or less) that it appears to be, the chance of finding one or more such individuals in almost any deliberative group would become a near certainty. Indeed, in large groups of about fifty, one should expect to find, on average, at least two or three such literates, while even in smaller groups numbering ten or so, the probability of having at least one literate approaches 90 percent. All this assumes randomly selected groups, of course, but unless the selection process is specifically biased against scientific literates, or against rationalists generally, as may be the case in some jury selections, these numbers should fairly approximate what would be found in practice.

The good news, of course, is that we really do not need *total* scientific literacy to profoundly alter the way that society deals with technical matters. Only about a 20 percent national literacy rate would achieve this objective. The bad news is that it appears we have no better prospect of achieving this literacy rate than the total (100 percent) literacy everyone agrees is impossible. It should be noted, by the way, that a scientific literacy rate of 20 percent is very nearly the same as the percentage of college graduates in the U.S. adult population (twenty-five years old or older). Hence, if all college graduates were literate in science, there would be no problem. Unfortunately, while they are a tempting target, we presently do not know how to motivate college students for such a formidable responsibility, any more than we do secondary school students, nor do we know what minimum education in science and technology is actually needed to prepare them for this role in society; and finally, we have no simple objective measures of such literacy. It would be far easier to increase the supply of scientists and engineers by 10, 20, or even 50 percent, if that were *really* needed, than to double or quadruple the scientific literacy rate of our nation, much as the latter might be desired. Considering that despite all of our efforts over the past half century or so we have not managed to budge the 5 percent literacy rate, this "20 percent solution," enticing as it may seem, is only an illusion.

Fantasy notwithstanding, however, it does add some useful perspective to the problem. Suppose society actually considered it *imperative* that 20 percent of the populace become literate in science, whatever the cost, for just the reason cited here, namely, to ensure that enlightened opinion is brought to bear on science-based social issues. If the science education community were charged with this task, how would it react? What student body would be targeted? Should it be the precollege level, where practically all the curriculum reform has been done to date, with discouraging results? Or should the focus be mainly on college students, where the chance of success might be better? It could not be simply a matter of designing new curricula, or of ensuring that all science teachers are thoroughly prepared in their subjects, the two obstacles most often cited as standing in the way of improved scientific literacy. These two factors alone probably could never achieve such a huge improvement in scientific literacy. Additionally, a very compelling form of incentive for students would have to be devised, possibly including a new college major leading to a joint B.A./M.S. degree and related professional opportunities, the master's degree in recognition of an agreed level of literacy in science. More will be said later on this point.

Obviously, this kind of analysis is not limited to the scientific literacy problem. It may be applied as well to determining the chance of finding one or more individuals having any special characteristic in a random group of N persons when the rate of occurrence of that particular characteristic in the general population is f. Thus, the individuals might be those who exhibit the habit of rational thought, or individuals who are not necessarily themselves literate in science but know how to get credible advice on science-based issues. In fact, we shall see that a possible solution to the scientific literacy dilemma may lie in developing appropriate machinery for providing the public with just such advice.

▲ SCIENCE APPRECIATION

Turning to the basic question of what the educated person should know about science, we should first discard our current notion of "scientific literacy" as being a meaningless goal and seek other criteria that may be more practical. Some that have been proposed may be worth considering. Testifying at a hearing of the Senate Armed Services Committee in November 1957 (soon after Sputnik was launched), the physicist and hydrogen bomb expert Edward Teller likened the need for public support of science to that of the arts. "Good drama," he said, "can develop only in a country where there is a good audience. In a democracy, particularly if the real sovereign, the people, expresses lack of interest in a subject, then that subject cannot flourish."[7] Later in the hearing, giving his views on education in science for the nonscience student, he added: "The mass of our children should be given something which may not be terribly strenuous but should be interesting, stimulating and amusing. They should be given science appreciation courses just as they are sometimes given music appreciation courses."

Teller's message of science *appreciation,* coming at a time when the American public, and particularly the Congress, was highly sensitive to the issue of Soviet competition in space, and just when massive NSF support for precollege science education was in its formative stage, fell on deaf ears as the nation girded itself for a far more ambitious role in science education, namely, to achieve in the educated public what had never before been accomplished—the intellectual state that came to be known as "scientific literacy." While not clearly defined at the

time (nor even now), this objective carried such a comforting pedagogical feel that one could hardly challenge its premise, and for the next
quarter century the science education community sought to justify virtually everything it did as bringing us closer to the goal of scientific
literacy. It tried valiantly but, as we have seen, it failed badly.

In our zeal to achieve the ideal we failed to question the goal itself
rather than the means of attaining it. The science education community was reluctant to concede the possibility that the answer might lie
in a goal other than scientific literacy, one that would be more acceptable to the general student and yet achieve a useful end result. The
science and engineering communities, and our nation generally, would
be better served by a society that, while perhaps illiterate in science in
the formal academic sense, at least is aware of what science is, how it
works, and its horizons and limitations. The general public is no more
sophisticated today in matters relating to science than it was fifty years
ago at the start of World War II—perhaps even less so because of the
rapid growth of science in that period and because the public now
tends to take so much of science for granted. Yet Teller's words may
have been prophetic, for here we are more than a generation later wondering, but not fully understanding, what happened to all the effort that
went into trying to achieve scientific literacy and launching still another crusade that is likely to fail as well.

If the science education community really understood what went
wrong with its previous campaign, it would realize that the major factors contributing to its failure have not changed. This is the reason why,
lacking a significant reform of the *purpose* of general education in science, the new effort seems destined to suffer the same fate. Meanwhile,
science and technology continue to grow and to impact all our lives, at
least indirectly. Research and development in these areas now account
for a little less than 3 percent of the Gross National Product, but the end
products resulting from such R&D consume roughly half our GNP. One
would think that so huge an enterprise would provoke more public
attention, which makes the problem all the more puzzling and the solution all the more difficult. The answer must lie in the fact that by and
large the public has become so nonchalant about technology as to give
little thought to how it works, or to its significance and problems.

How might one go about cultivating the science appreciation advocated by Teller, who unfortunately gave no advice on this crucial
point? Obviously not through the usual courses in science, for we know
this will not work. Could it be done through special courses in science
appreciation, just as we have courses in music and art appreciation?

Perhaps—if a real effort were made to develop an effective curriculum toward this end and one had a clear idea of what should be meant by science appreciation. Teller probably made a minor tactical error by comparing science appreciation with art or music, for the latter have a more casual or emotional connotation than the sort of understanding we probably want in science. Also, "appreciation" has the twofold connotation of gratitude and awareness, which may be somewhat confusing. Perhaps a better term, because of this allusion, would simply be "science awareness," which we shall use later on in describing our proposed goal for general education in science.

Almost fifty years ago, the approach taken by James B. Conant in his well-known Harvard Case Histories in Experimental Science[8] marked a significant effort in this direction. Its premise, however, that appreciation of science might grow out of the historical development of some of the great ideas in science, evidently did not fully satisfy the need, as seen more in the eyes of Harvard faculty than by its students. Nor, as we shall see, would it fully satisfy the needs of science awareness, although a strong case can be made for it as part of any science curriculum. The main advantage of a historical approach is that the science it deals with, when viewed from a modern perspective, is generally much simpler and the reasoning involved is likely to be more transparent than in a contemporary example. However, it would be inconceivable today, largely because of its societal import, to omit contemporary science from a student's formal education. Thus any modern science curriculum, even one taking the case history approach to demonstrate the process of science, must also lean heavily on contemporary science, particularly on some of the scientific advances that gave rise to modern technology and its consequent social issues.

So, there remains the major question of how one should deal with contemporary science, for experience tells us that a heavy concentration on science content will not work at all with the general student. Here the emphasis should be on those topics, as suggested earlier, that influence public attitudes on science and affect the individual most directly.

▲ THE PURPOSE OF GENERAL EDUCATION IN SCIENCE: A NEW LOOK

What should be the purpose of science education for the general student? It is clearly unreasonable, in the short time that such students spend with science, to expect them to see the beauty in science that Henri Poincaré spoke of, the beauty that resides in "understanding the order of the parts."[9] Nor is it reasonable, as we have seen, to expect the general student to attain a level of understanding in science that would permit the future adult to reach truly independent judgments on societal issues having a scientific base. Hence society must learn to rely on scientific experts for advice, as it does in other highly specialized fields such as law and medicine, and students would be best served by providing them with useful guidelines on how to select such experts. Nor, unfortunately, can we hope to develop in students the "scientific habits of the mind" so strongly championed by Dewey as an outcome of training in science; but we can hope to show the power of rational thought *as it applies to science.* Perhaps there will be some fallout affecting the individual's everyday habits of thought, but we should not count on it. The Nobel physicist I. I. Rabi, commenting on the need for truth and clarity in the practice of science, asked the rhetorical question: "Are we living in a society where truth and clarity are an important element of our culture?" And his answer was that "we continue to ignore clarity, logic and even arithmetic."[10]

Teller was perfectly correct in his observation that science must have an appreciative audience, meaning in these times a supportive society, one that values science for its intellectual strength as well as the foundation for the technologies it spawns. Without such support, science and technology, activities that together account for such an appreciable fraction of our GNP, could both founder, and the United States might indeed become a second-rate nation, as some of the more dire forecasters predict. It is often said that the scientific community speaks with two different voices when it urges public understanding of science. One is the voice of reason, an intellectual appeal to the effect that (a) because science has become so pervasive in modern society it warrants everyone's attention, if not active interest, and (b) because it represents a special mode of inquiry that has proved successful, it should be stimulating to all educated minds. The other voice is more pragmatic, and says that by urging public understanding of science it is really seeking approbation, or at least appreciation, of what scientists and engineers do. Some consider this merely a self-serving appeal; others contend that it is important not only to the welfare of the scientific enterprise but also to society itself that the public approve (or disapprove in some instances) of what the enterprise does, assuming that its conclusions are based on reason rather than emotion.

Thus, one objective of general education in science, perhaps the most important one in the eyes of the scientific community, must be to encourage such an appreciative audience, one that at least understands how and why so much needs to be spent on science and technology, even apart from military requirements, to keep pace with the developing world. Another objective, probably more important to society at large, must be to help the student, and society generally, feel more comfortable with new developments in science and technology. They need not so much to understand the details but to recognize the benefits—and the possible risks. In principle, this could be accomplished in two ways. The obvious way, of course, is by developing in the student true (and lasting) scientific literacy, which, we have seen, is an impossible task for all but a very small fraction of the population. The other is by helping the public to gain confidence in the individuals, public interest groups, and governmental agencies that control the funding of science and technology, and in an important sense regulate the interface between technology and society.

▲ EXPERT ADVICE (AND THE IRRATIONALISTS)

Having concluded that achieving widespread scientific literacy, in the sense of empowering the individual to reach wholly independent judgments on science/technology-based social issues, is out of the question in the foreseeable future, the only practical solution is to provide the public with easy access to responsible, expert advice on such matters, just as we do in other specialized areas such as law, medicine, economics, government, politics, and so on. It may be that the real mark of scientific literacy in the general public, as suggested by James Conant,[11] is not how much science the individual knows but how well that person can seek out and filter "expert" scientific opinion (see chapter 4).

The use of expert opinion is a sensitive subject, perhaps more so in science than in most other areas of human activity. Why this should be so appears to be linked to the use that the individual makes of professional advice and the popular belief that scientific judgments ought to be black or white. Individuals generally seek professional advice because they have personal problems. We go to physicians because we have medical problems, or to lawyers to resolve legal problems, or to marriage counselors or business and investment advisors, and we sometimes seek second opinions because any decisions we may reach based on such advice are very important to us personally—may, in fact, involve matters of life and death or one's personal liberty. Hence, such interactions are characterized by (a) an opportunity for personal appraisal and determination of trust in the expert, and (b) a *personal* stake in the outcome. How do we select such experts? Sometimes through hearsay, more often through referrals by friends or business associates, or, too infrequently, by inquiring into evidence of success or reputation of the expert among his or her peers or appropriate accrediting agencies. Questions of bias or conflicts of interest rarely enter our minds when dealing with such experts, or if they do in some instances at least the individual has an opportunity to explore these questions directly with the expert. What we generally seek is expert *status* rather than the more important quality, expert *ability*. The former ought to include the latter but sometimes does not. For example, pseudoscientists, frauds, and charlatans often gain "expert status" in the eyes of their followers by deception rather than through professional competence.[12]

The situation with science advice is somewhat different. First, such advice generally does not involve the individual personally but is provided mainly to influence public opinion on societal issues having a scientific or, more often, a technological base. Second, unless the individual is a member of some local action group with the opportunity to interact directly with such experts, the advice generally comes secondhand through reports or statements appearing in the mass media. In this respect one might think it no different from "expert" political advice at election time, for example, except for one feature. The public expects political experts to be biased, but not scientists. After all, doesn't science deal with hard facts, with logical reasoning and unequivocal judgments?

With the first two of these, yes, but different scientists, all experts technically, may interpret the facts differently, not only when dealing with purely scientific matters, but particularly in their relation to social issues. In science, such differences are usually resolved through technical debate and appropriate experiments. In the social arena, where the "science-based" issues generally have to do with personal health and safety or with ecological concerns, and where they often deal with barely detectable marginal effects, it is rarely possible to carry out decisive experiments. Thus the experts often have to extrapolate from limited or unrelated experience, or can only speculate on the societal impact of a given technology. This sort of uncertainty understandably gives rise to differences of opinion among competent scientists, which the public, in its overinflated image of scientific authority, often takes to mean that science cannot be trusted to deal properly with related social issues. In the absence of such trust, the public is easily swayed by pseudoscientific opinions, impostors, and self-serving demagogues posing as defenders of the public interest.

The media may be fascinated by such controversy but the public is understandably confused by opposing opinions in science. The neat package it expects from science becomes undone and the public is left up in the air. The fact is, however, that educated judgments, even though speculative, are better than pure guesswork or inferences based on misleading advice, emotion, or personal bias, which are the only other avenues open to the public. Hence the prudent choice is to seek the best scientific advice available and to sort out differences in the way we always do in areas that we know little about, namely by choosing advisers on the basis of their reputations and the confidence they inspire in us.

There are many who condemn this means of seeking scientific advice on social matters on the ground that it disenfranchises the individual who may be called upon to act or vote on science-based societal issues. Their arguments often strike a sympathetic chord, especially among college students. Among others, both the late antitechnologist Jacques Ellul,[13] a French Protestant theologian, and the social critic Lewis Mumford[14] have warned of the dehumanizing potential of technology. Ellul goes further to deplore the danger of society being dominated by rationality and logic to the exclusion of spontaneity and creativity, while Mumford appears content only to caution that if the individual becomes subordinate to automation, his confidence in making independent judgments could be undermined. From these it is but a small step to postulate that reliance upon expert opinion on technology issues is a part of this dehumanizing process and may represent a danger to democracy. Neither writer presents convincing, and certainly not objective, evidence for his concerns; in fact, if one takes Ellul's views at face value, objectivity would be equated with rationality and should therefore be suspect over emotion and irrationality as a reliable guide to truth. Ellul's writings found a receptive audience among politically disillusioned Americans during the 1960s, particularly on college campuses, but reasonable critics of his social theories rose to the attack. Notable among these was Alvin Toffler, who characterized Ellul as a "French religious mystic" and one of the most extreme of "a generation of future haters and technophobes."[15] Going to the extreme, the counterculturist Theodore Roszak, in his critical analysis of American culture, provided an even more caustic view of expert advice:

> Does our democracy not continue to be a spectator sport in which the general public chooses up sides among contending groups of experts, looking on stupidly as the specialists exchange the facts and figures, debate the esoteric details, challenge one another's statistics, and question one another's prognostications? It is difficult to see that, in the long run, such a counter-balancing of expertise can be a real victory for the democratic autonomy of ordinary citizens. They remain expert-dependent.[16]

Philosopher Paul Feyerabend takes a similar position regarding reliance upon experts, asserting that one should not accept the judgment of scientists or other specialists without subjecting it to "the most painstaking scrutiny," even to the extent of establishing committees of laymen to decide whether a given scientific theory such as evolution should be considered valid.[17] Pointing out that even experts make errors, and that the discovery of such errors by ordinary people is the underlying assumption of trial by jury, Feyerabend concludes that the average person can acquire, through hard work, the necessary knowledge to challenge the experts, and hence "science is not beyond the reach of the natural shrewdness of the human race." What Feyerabend fails to point out, however, is that jurors usually do not understand the technical details of expert opinion; nor do most judges. Even lawyers often fail to grasp the subtleties involved, and the art of cross-examining *competent* experts often boils down to creating confusion in the jury's mind between the possible and the probable. One could not quarrel with the proposition that the ultimate answer is to ensure that everyone becomes sufficiently literate in science and technology so they can reach independent judgments on all such matters. Yes, by working *very* hard at it the average individual might acquire sufficient insight to challenge the experts on a particular scientific subject. But this is so specious an argument, given the realities of learning this much science, as to defeat his entire premise. Feyerabend's argument is actually little more than posturing. Does "power to the people" really mean casting aside common sense, as well as reason?

In a sense, it is pointless to debate Roszak or other counterculturists on the role of science and technology, for by deliberately cloaking themselves in the mantle of irrationality they render the playing field very uneven. Such "irrationalists," which is perhaps a more descriptive term than "antiscientist" or "counterculturist," seem to realize their self-imposed invincibility and take full advantage of it. How could the science education community, whose guiding principle is the use of reasoned argument in science, agree to abandon rationality as the starting point of any dialogue, social or otherwise? The fact that one's everyday thoughts and actions are often guided more by emotion than by reason does not mean that we should yield our faith, as Judge Learned Hand put it, in "the eventual supremacy of reason."[18] To do otherwise would be inviting intellectual anarchy. It may well be, as scientist Roald Hoffman contends, that the rational world of the scientist is often ill suited to the resolution of political or societal problems, for scientists are at their best when "motivated to speak as the voice of reason: to give sound advice, to counter ascendant irrationality."[19] Most counterculturists would have us give up on rationality and on the belief that science offers a means of studying nature *objectively,* both of which are rejected by those sociologists and fringe science groups who would like nothing better than to insinuate their worldview into, and gain control of, such educational departures as the STS movement. In a sharp rebuke to the professional irrationalists, the late philosopher Charles Frankel analyzed their objectives:[20]

> As a pragmatic matter, what irrationalism asks is that society invest less—or nothing at all—in maintaining the institutions, and the code of ethics and etiquette, which have proved necessary to support the emotion of reason. . . . In brief, for the irrationalist, the universe is good; it is man, rational man, who has willfully made it evil, all by himself. Irrationalism, behind its long arguments and often impenetrable rhetoric, is an attempt to solve the ancient problem of Evil and to restate the ancient myth of the Fall.

And in pointing out how reasonable objections to some of the abuses of technology can breed unreasonable generalizations of the scientific enterprise, Frankel went on to say: "The careful rational methods by which knowledge and technique have been advanced have only rarely been used to examine the purposes to which this knowledge and intelligence are harnessed. It is natural that science, in such a setting, should seem to be a Frankenstein to those who are threatened by it."

The irrationalists among the social scientists are not so much interested in science as such, but in cutting it down to their own "size." The answer to this should be abundantly clear to all science educators: whatever the curriculum or style of teaching, respect for the rule of reason, for truth, and for disciplined honesty must predominate, for these are the cornerstones of science, as they should be for all *responsible* behavior. If general education in science is to have any meaning at all, at the very least it should help students to distinguish fact from fiction, science from pseudoscience, and rational arguments from irrational. Science education, whatever its future course, must not become a platform for the nonsense dispensed by the neo-Luddites and other anti-science/anti-technology movements. Instead, the science class-room, and particularly that portion of it devoted to such multidisciplin-ary topics as STS, should be the forum for debunking the attempts of such fringe elements to distort the public mind, first by exposing their tactics, and then by stressing over and over again the central role in science of *objective, reproducible* evidence.

One can seriously question whether, in fact, technology is more of a threat to the democratic process than the irrational solutions of-fered by the anti-technologists and counterculturists, which, in part, would demolish many of the technologies we presently have and re-place them with simpler, easier to understand systems (see chapter 5). Beyond this, which is the *real* threat to democracy—reasoned judg-ment or the irrational conduct of the science counterculturists? Not only are they largely ignorant of science, which is understandable, but also, apparently, of the social forces that guide the development of technology, which is unforgivable. They seem to live in a social never-never land as far as technology is concerned, where reality is ignored and fantasy becomes the rule. How else can one judge a philosophy based on irrationality? What criteria can be applied to statements hav-ing no rational basis? Indeed, one is sometimes tempted to ask the irra-tionalists whether it is rational to believe in the irrational, thereby setting off still another philosophical storm. The absurdity of their posi-tion should be obvious. Rationality takes some conscious effort; the irrational requires none. Raging at technology will not bring it to a halt, any more than would commanding the Earth to stand still. Just as civili-zation always has managed to cope with the new, it will learn to live with the inexorable growth of technology, and to do this it must either learn to understand it or to rely upon expert advice.

▲   SCIENCE EXPERTS AND THE "SCIENCE COURT"

The only sensible solution to the problem lies in the judicious use of scientific experts, not as surrogates for the public in determining the proper course of action on science/technology-based social issues, but as advisors on the purely technical aspects of such issues, from which the public might hopefully reach better-informed judgments. The problem is that in most cases experts have been called into adversarial, even confrontational, proceedings, often having political agendas as well— whether these be legal matters, community issues, or legislative hearings—and with opposing sides already drawn there is little chance of resolving the scientific issues among the experts beforehand. If there were an opportunity for the scientists to seek a common understanding before stating their positions publicly, many of the perceived differences might be settled early on. At the very least, one should expect that where differences still remain these could be spelled out in plain enough language so that nonexperts can make the final choice based on reasoned alternatives.

The "Science Court" concept proposed by physicist Arthur Kantrowitz a quarter century ago offered a novel and imaginative means of resolving the technology dilemma confronting the public in controversial policy matters.[21] Basically, this proposal envisioned a formal courtroomlike adversary procedure in which, after separating the purely scientific from the moral and political components of a given controversy, panels of technically competent scientist/advocates and judges, rather than the uninformed public, would establish the factual components on which subsequent political decisions might be based. It was believed that such a quasi-judicial procedure could better resolve conflicting scientific claims than the usual confrontational system employed in courtrooms or other adversary hearings. How the results of such procedures would be used, whether advisory or as binding on the parties involved, was not specified but left to the prior agreement of the courts or hearing officers and the litigants. Also left vague were such important details as how and when to initiate such Science Court procedures, how to select the judges and advocates, and how to make sure that the proceedings are not dominated by the traditional conservatism of the scientific establishment.

Nevertheless, as a means of separating fact from fiction, rumor, speculation, or emotion, the Kantrowitz proposal seemed to many at the time a promising start on a perplexing problem, and even prompted a few serious experiments, mostly in government agencies. Regrettably, however, while remnants of the original Science Court idea are still practiced occasionally, generally with cross-examinations of scientists being carried out *in camera,* the concept soon floundered and died for lack of strong support outside the science community. The legal profession opposed it for various reasons. Lawyers were reluctant to give up their traditional roles in the courtroom, arguing obliquely that the proposed Science Court could not achieve its stated goals because it tended to oversimplify a highly complex process, namely, seeking to resolve issues by reaching consensus through an adversary procedure, that is, having both sides satisfied with the outcome. Every scientist knows that particularly at the cutting edge of science, prevailing scientific opinion can change over time, and every good trial lawyer knows how to take advantage of this in the courtroom by challenging the expert's lack of absolute certainty. Everyone connected with the legal system is aware that "junk science" is sometimes practiced in the courtroom, especially in tort actions, both by unscrupulous maverick scientists qualified as experts by judges who lack reliable methods of distinguishing the quacks from true experts, and by lawyers who regard their chief function as discrediting opposing experts and creating confusion and uncertainty in the minds of jurors or hearing panels.[22]

But the legal profession was not alone in opposing the proposed Science Court. So too did many sociologists and most citizens' groups, both objecting that such a reductivist approach tended to emphasize facts over social values, as though facts were not also value laden, thereby deflecting the public's criticism of technology. The sociologist Dorothy Nelkin, in summarizing the many social and political objections facing the Science Court idea at the time, laid bare the core of the opposition by stating: "In sum, the effort to resolve factual questions apart from questions of value is exceptionally difficult and, in the heated context of political controversy, may not be very useful. The point of disagreement in conflicts involving scientific information seldom lies simply in questions of scientific fact."[23] And in a subsequent paper with Sheila Jasanoff, Nelkin struck again at the notion of value-free knowledge with the statement: "The belief that scientific expertise is inherently removed from value considerations and that scientists are therefore political celibates is an anachronistic and even dangerous one."[24] Indeed, Jasanoff and Nelkin seem opposed to any means of helping judge and jury to better understand the factual science component of litigation: "Equipping the courts with scientific and technical support may facilitate the adjudication of these issues; however, it may also divert attention from the public responsibility for major policy decisions and encourage the conversion of moral and political questions into technical debates among experts."[25] The context, of course, was that since such scientific expertise is invariably intended to influence policy decisions, it cannot be considered apolitical or impartial. However unreasonable such hypothetical objections may seem to most scientists, it must be recognized that some segments of our society would prefer to have the public flounder and muddle through in ignorance of scientific matters rather than accept the possibility of value-free knowledge, even in science, or agree that highly technical issues might be better resolved out of the public forum.

Among those who hold that public debates can eventually resolve such issues, one senses an underlying belief that direct confrontation of "experts" by the uninformed public is the best *means* of reaching an answer to a policy question involving science or technology, even if it may not the best *answer*. That is, even if the public does not understand the technology involved, the opportunity to air its views and to challenge the experts provides comfort that its commonsense judgments are correct. In other words, consensus, even though it may be arrived at incorrectly, would appear, under this theory, to be more important than truth. But while a given use of technology may properly be decided by popular vote, it is folly to believe that reasonable scientific conclusions can also be reached by this means.

Still another objection to the Science Court concept when it was proposed was that several advisory bodies already existed, notably the ad hoc commissions of the National Academy of Sciences, which adequately served the purpose of informing Congress and the public on science and technology issues. This would be disputed by many in the scientific community, and in government agencies as well, who do not believe that truly independent, impartial advisory bodies currently perform this function well. In fact, Kantrowitz maintains that the National Academy, which is generally looked upon as the chief arbiter of scientific issues in the United States and is largely funded by Congress, has a conflict of interest when it appoints fact-finding commissions that outwardly appear to take this function out of the hands of Congress and the executive branch.[26] Many others agree with this view of the National Academy, more as a matter of appearance than any factual concern over its impartiality.

The Kantrowitz proposal, or some reasonable modification thereof, has not been given the fair probation it deserves. Granted that finding the right formula is not easy, the use of qualified experts to facilitate the process of conflict resolution in the science/technology arena offers the only *practical* solution to the vexing problem of scientific illiteracy in the mass of our population. There can be little doubt that because the problem has become so pervasive, science-based litigation will one day be sheltered under such an umbrella; and when those who presently oppose the use of scientific experts realize that widespread scientific literacy is not around the corner, public policy debates will also seek out expert advice. Seeking better ways of handling scientific issues involved in litigation is already under way. A recent report by the Task Force on Judicial and Regulatory Decision Making of the Carnegie Commission recommends, among other steps, that judges assume a more active role in managing how science and technology issues are presented during litigation, and that institutional links be established between the judicial and scientific communities.[27] And in a 1993 decision, the U.S. Supreme Court called on judges to take a more active role in keeping "junk science" out of their courtrooms. We can therefore expect judges to set up their own expert panels to advise the courts on the scientific aspects of any litigation. While not quite the solution proposed by Kantrowitz, it is clearly a step in that direction, the end result being that more reliance is placed upon credible experts in science.

If the American people, or those individuals and groups who undertake to speak for them on such matters, really believe it is worthwhile to find ways for the public to participate more fully in societal issues based on science and technology, a mechanism must be developed to seek out and act upon credible expert advice. After all, society has learned to rely on experts for advice in other areas of personal need that may also involve value issues; why should specialized knowledge in science and technology be regarded so differently? One can hope that reason may yet prevail, and that public policy debates will move to a system incorporating some form of Science Court concept, or something akin to it, as outlined below. The alternative is to give up all hope, as did Roszak, that true, unbiased expertise can ever be brought to bear on socioeconomic problems involving technology, and resign ourselves to a future of uncontrollable technology determined solely by a technological elite, guided only by what he called "non-intellective thinking."[28]

▲ A NATIONAL SCIENCE WATCH?

Most people would agree that some way should be found to provide the public with sound advice on science-based issues, a reputable (and credible) forum that can help guide the public to informed choices rather than leave it at the peril of well-organized vested interest groups or pseudoscientific frauds. Developing such a vital interchange would serve not only the best interest of the public at large, but that of the science/technology community as well. To achieve this, both the public and the scientific community will have to work together to bring about a system of providing the best possible advice to society on science and technology issues. There are many public interest groups and governmental agencies, we know, that offer such advice. It does not appear, however, that any of these inspire the full confidence of the public, for they are generally not regarded as being totally free of bias or private agendas. An environmentalist group offering expert opinion on the effects of pollution or on a theory of global warming, government agencies offering contrary opinions on the same subjects, an animal rights organization insisting that the biomedical sciences can find other ways besides animal experimentation to do its research, social reform groups, professional science and engineering organizations; all are somehow suspect, sometimes unfairly, but more often with good reason, of advancing their own special interests rather than providing objective, unbiased scientific opinion.

It is hard to dispel the perception that self-appointed public interest groups or government agencies invariably have private agendas that may differ from their public posture, and there appears to be a widespread mistrust of scientists as advisors to the public. Table 8.4 shows some interesting results of a study on public knowledge of science conducted by Schibeci in Australia.[29] Note that Australian adults hold a somewhat more realistic view than Americans of their shortcomings in science, but, in common with most personal surveys, they still believe that their own knowledge is better than that of their neighbors'. The test sample, incidentally, did not have a university education, but most (62.5 percent) had gone beyond the minimum compulsory level of school year 10. According to Schibeci, roughly the same results have been found in the U.K. The most telling feature of this study is found in the last test item of table 8.4, and in responses to other questions dealing with public perception of science and scientists. What we can conclude from the study is that most of the respondents were bewildered by what they perceive as "infighting" within the scientific community and tend to view scientists as uncaring about the future of the world. Also evident from the responses is the power of the press and public relations in shaping the image of science and scientists: most respondents believed that scientists are more believable when they are well known, or work for well-known organizations. Whether these perceptions change much with higher education is unclear, but the indications are that they do not.

▲ **Table 8.4** Public Perceptions of Its Knowledge of Science and Technology

How well informed would you say most West Australian adults are in the general area of science and technology? Would you say . . .

Very well	0.8%
Fairly well	17.6
Neither well nor poorly	19.1
Not very well	48.4
Not at all well	12.5
Don't know	1.6

And how well informed would you say you yourself are?

Very well	0.8%
Fairly well	30.1
Neither well nor poorly	22.7
Not very well	32.8
Not at all well	4.7
Don't know	0.8

Generally do you think that science and technology do . . .

More good than harm	58.2
Equal good and harm	34.0
More harm than good	5.1
Don't know	2.7

SOURCE: R. A. Schibeci, "Public Knowledge and Perceptions of Science and Technology," *The Bulletin of Science, Technology, and Society* 10, no. 2 (1990): 86–92 (table 2). Reproduced by permission.

To achieve a better accord between science and society, a new kind of organization is needed, something in the nature of a national "Science Watch Committee," whose function would be to monitor activities in science and technology and to keep the public informed through the media of the improper use of existing technologies and of the meaning and significance of new technologies, including, whenever feasible, objective risk-benefit assessments. As with the Science Court concept, the problems facing such a proposal are enormous but not insurmountable if the public interest groups are willing to give it a fair trial, at least through the planning stage. Probably more than one such committee would be needed to cover the full spectrum of issues at the science/technology–society interface. And perhaps this should be a global undertaking, with separate national watches operating under an international umbrella, since so many environmental and ecological problems transcend national borders. International treaties on such matters, formulated by government agencies, cannot be totally free of political and economic factors; nor can the many nongovernmental groups with preconceived briefs claim to be free of social or political bias. It is clearly impossible to demand that members of such watch committees be totally free of bias, but it would be prudent to require that their involvement in committee activities be strictly limited to the technical issues involved, and that the committee members at least have no *recognized* political or social agendas.

Such science watches would ideally consist of members nominated by respected scientific organizations, including the social sciences, and perhaps some labor leaders and representatives of concerned citizens' groups. Probably the scientist members should be university based to ensure their independence, perhaps taking leaves of absence for a year or two to serve on a science watch committee. In fact, it may be that a consortium of universities would form the ideal structure for such an undertaking. The committees must be constituted, financed, and institutionalized in such a manner as to reflect in the public mind an aura of expertise and independence, to the extent that the popular media would willingly abstract committee reports and hopefully seek committee reactions before reporting on unfounded claims and rhetoric of the many individuals and organizations that offer gratuitous advice to the public on sensitive technical issues. And hopefully, because such committees would be considered credible and impartial, legislators would eagerly look to them to balance the claims of lobbyists who seek to sway them on pending or needed legislation.

This is less "blue sky" than it might appear, even though we are dealing here with issues that are often heavily burdened with emotion, and with experts who are nonetheless individuals living in the real world and who cannot be expected to divorce themselves completely from the normal flow of everyday life. However, as Kantrowitz pointed out in his Science Court proposal, it should be possible in such situations, particularly where the committee is free of political and economic constraints, to phrase the issues in such a manner as to separate the purely technical from the societal questions. This is not to suggest that scientists are free of personal bias; no one really can be, not even, as we all know, can justices of the U.S. Supreme Court. To believe otherwise is wishful thinking. Why should a society, all members of which are biased in some respects, expect to find in its midst some science experts who are otherwise intellectually neutral, who have no opinions whatever on social, moral, or ethical questions, only on technical matters? All one can reasonably expect is that the experts selected are free of obvious conflicts of interest or any other personal commitments that might limit their ability to provide completely *independent* opinions on the issues put before them.

We cannot hope to find scientists who are totally free of personal bias on social matters; that would be unreasonable to expect and impossible to determine. But we can expect to find scientists who are free of technical bias, which is relatively easy to establish. In most instances, technical opinions can be value free; they become value laden only when society seeks to take action on them. The mistake made by many social scientists is to conclude that because in the end most science-based societal issues are burdened with value judgments, some very heavily, even the science involved cannot be value free. But natural science has easily resisted the constant efforts of counterculturists and social revisionists to force it into the social mainstream. They try to show, often by tortuous reasoning, sometimes by pure conjecture, that even the practice of science can never be free of personal bias. But they are badly mistaken, for they simply do not understand the way science works. What can be expected from a group of scientific experts, working together in as free an environment as possible, is in most cases at least a consensus, if not total agreement, on the purely *technical* issues involved, plus reports to the public that clearly explain their differences, where such appear.

The big question, of course, is how to organize such watches as to generate public trust, and this requires a great deal of careful thought and planning. That something of this kind must come about eventually is almost certain, for regardless of how much one might deplore the need for expert advice on science and technology, there is no practical alternative. A logical first step would be to convene a study group with private foundation support to determine the best means of organizing and funding such science watches. If the social science community were to get behind such an experiment, it would move ahead much more quickly and be more likely to reach a meaningful conclusion.

CHAPTER 9

Conclusions—and the Road to a Possible Solution

It is as fatal as it is cowardly to blink facts because they are not to our taste.
—John Tyndall (*Fragments of Science*, Vol. II, 1896)

▲ By now it should be apparent that the notion of developing a significant scientific literacy in the general public, as we have come to understand its normal meaning, is little more than a romantic idea, a dream that has little bearing on reality. We have seen that while some of the arguments favoring such literacy may be valid, most are impractical, particularly as the American public is not convinced that the boost such literacy may give to their quality of life is worth the energy expended. We have sought to make clear on two separate grounds that scientific literacy is not merely a matter of formal education: first, despite the ever popular school bashing, the problem does not lie solely (or even mainly) in our schools; curriculum reform and better teacher preparation *alone* cannot provide the needed stimulus for a more literate society, whether in science or any other academic discipline. That must come from outside the school setting, from widespread acceptance by the public of the idea that becoming and remaining literate in science may actually be in one's self-interest, and here, we have seen, is where the problem really lies.

Second, we have shown that equating school literacy with adult literacy in science, where it might actually count in a societal sense, is a fallacy; the fact is that whatever we may seem to accomplish with students that somehow relates to scientific literacy appears to have little bearing on their perceptions of science as adults. Becoming literate in science, however one chooses to define such literacy, is one thing; remaining literate is quite another. To further confuse the issue we have seen that public literacy in science means different things to the various "publics" in our society, depending on their incentives and perceived needs to be informed in science.

▲ SCIENCE AWARENESS: A NEW SCIENTIFIC LITERACY

Having delineated the many problems confronting the scientific literacy movement, we now address the question of how to reformulate the goal itself. After all these years of trying to achieve scientific literacy in the student population we have made no discernible progress. Everyone seems to have a vague, ill-defined notion of what *should* be, but we've had no clear specifics, least of all on how to achieve our goals. All is not lost, however. The fact that there is no general acceptance of what the elusive scientific literacy actually means, or should mean, is itself significant, for it forces us to reflect on why we think scientific literacy is important, whatever one takes it to mean. Since the curriculum, in effect, sets the tone of a science program, on what should curriculum developers (and textbook writers) concentrate in any new approach to science education?

Putting aside science as a cultural imperative, there are two valid reasons for seeking widespread public understanding of science—three, if one includes the practical but unspoken objective of getting critical masses of students into our science classes. The first is the desire of the science and science education communities to have the public (and particularly public officials) appreciate and support these enterprises to ensure their continued vitality. This purpose might better be classed under the science or technology *appreciation* alluded to earlier. The second is the desire, mainly of the social science community, to have the public participate directly in the decision-making process on societal issues having a scientific base, issues that are actually based for the most part on technology, as we have seen.

This is what most people take scientific literacy to mean, and is where the bottleneck actually occurs. Prodded mainly by the social science community, we have been laboring to find ways of educating all Americans in science to the point where they can reach *independent* judgments on such issues. Such *social* or *civic literacy* is really the underlying goal of the scientific literacy movement, and particularly its STS (science/technology/society) component, and it is time we all recognize this as an impossible task and get on with the normal business of science education. Not even professional scientists can always be relied upon to vote with their heads instead of their feet, and no *reasonable* amount of science education can ever get the average person to the point where he or she is able to judge such issues independently and dispassionately.

The answer is simple and straightforward, if one takes as a given that the only practical way of resolving the primarily technical aspects of a given issue is to seek the advice of experts in the field. With this assumption, the three guiding principles for presenting science to the general (nonscience) student should be:

1. Teach science mainly to develop appreciation and awareness of the enterprise, that is, as a *cultural* imperative, and not primarily for content. Designing curricula aimed at this objective is not difficult, provided one does not attempt to cover everything.
2. To provide a central theme, focus on technology as a *practical* imperative for the individual's personal health and safety, and on an awareness of both the natural and man-made environments. Several very good materials along these lines already exist.
3. For developing social (civic) literacy, emphasize the *proper* use of scientific experts, an emerging field that has not yet penetrated the science curriculum.

The last of these may not be so easy, as we have no experience with it on a large scale, yet it may be, as we have seen, the only valid measure of scientific literacy. There is some strong opposition to this approach, we have seen, mainly from the social science community, which generally believes that relying upon experts in such matters disenfranchises society. This might be a valid argument if the only practical alternative, namely, maintaining the status quo, were acceptable. But it is not, as almost everyone agrees. Hence there is the ongoing crusade for full-scale scientific literacy—for everyone to be his or her own scientific expert—a mission that has thus far failed painfully and is doomed to continued failure in the future. Surely, if the social science community really wished to do so, it could devise means of seeking out reliable, unbiased expert advice on the technical side of science-related issues without compromising the independence of the individual in the decision-making process. After all, providing the public with the means of acquiring information essential to rational decision-making in a democratic society should be the responsibility of all educators, not only those in science. One such proposed course at the university level was described recently by William Rifkin, then a science Graduate Fellow in the Program in Values, Technology, Science and Society at Stanford.[1] All curriculum developers, especially those seeking to incorporate STS segments into their programs, should give serious thought to the subject of science advice.

▲ ACHIEVING SCIENCE AWARENESS

Science is invariably presented to students, and to the public, ex post facto, so to speak, as a fait accompli, with facts, theories, and laws already tied into a neat package to be opened, examined, and admired (or scorned) by the prospective learner. Indeed, it cannot be done otherwise for the general student, who spends so very little time with formal science—in many cases, unwillingly. The problem is, however, that beginning in the secondary schools, instead of being wrapped in a single package science comes in a number of packages, each one a part of science yet containing the artifacts of separate and seemingly discrete disciplines—biology, chemistry, physics, earth science, astronomy—or even subspecialties of these. Thus the main impression the student gets is that science is highly fragmented, yet we still hope that the learner will somehow see the forest instead of the individual trees. Where is the common element called "science"? In some cases we find the discipline divided into only two major categories, physical and life sciences, and occasionally courses are offered under the generic heading of "general science." But these are merely course titles, and within such courses students soon find the subdivisions that characterize the partitioning of modern science. The point is that our science education focuses on courses rather than on science or on students.

What we really want the student to be aware of, rather than simply the aftermath of the process, is science in the making: what goes into preparing these packages, what science does, and how and why it goes about its practice. It seems that only in the elementary schools is there still a *science* curriculum; everywhere else, influenced mainly by the pressure of academic scientists, science has been broken down into its separate disciplines. There are good reasons for this, of course, but they are professional rather than pedagogical. While the practice may seem sensible to the serious science-bound student, it not only creates confusion in the minds of the nonscience students who make up the vast majority of the precollege student population, but also makes it impossible to teach the truly important aspects of science, namely, the nature of the enterprise itself.

Interestingly, in the "alphabet soup" days of curriculum reform (during the 1960s and 1970s) it was recognized that in the elementary grades science process and the conceptual framework of science were more important than science content. Several notable programs were produced, only to be lost sight of in recent times as new programs were developed to be "teacher-friendly" because the earlier ones demanded too much of the average elementary school teacher. But instead of solving the problem by providing elementary science specialists in all the schools, which would have been costly, or by seeking other ways of bringing competent science support into the elementary schools, granting agencies and book publishers turned to the time-proven method of making the curriculum teacher-proof. And strangely, with all the curriculum reform that took place at the secondary school and college levels, science content continued to dominate the curriculum. For some time there has been a movement in the United States seeking common elements among the sciences to form the basis for a unified science curriculum,[2] but to date it has had little effect on the standard science curriculum, largely because it is considered too radical a departure from accepted practice. Yet this is just the sort of approach one should take in order to impart more of the nature of the discipline to the general student. Unified science should be distinguished here from another active movement known as "integrated science," which is an attempt to present science in the context of other disciplines such as history, philosophy, or social studies.

What we seek is a society that (a) is aware of how and why the scientific enterprise works and of its own role in that activity, and (b) feels more comfortable than it presently does with science and technology. Hence our main objective should be to provide students with an *accurate* image of science that may serve them in the future as productive members of society. Remember that, as we have repeatedly emphasized, having literate students does not necessarily mean that they remain literate as adults—the staying power of conventional science courses is notoriously poor—and it is only the adult population that is in a position to apply its awareness of science to the good of society. Hence, in the belief that understanding something about the nature of science would be more useful to everyone than learning (and then forgetting) a collection of facts about nature that more properly belong in the category of natural history, we should be teaching what the scientific enterprise is *about* rather than presenting the packaged science content that is characteristic of virtually all introductory courses.

We begin by asking just what it is about science that we would like the average person to understand, or at least be aware of. We first remove from consideration the simple *facts* of science, the sort of knowledge to which everyone is exposed in the course of living and schooling—the sort we have been calling "natural history." Not that these are unimportant, but they are details that one learns by combined rote and experience: simple astronomical and environmental knowledge, biology as it relates to one's health and safety, and everyday ("kitchen") chemistry and physics. Among these are a practical knowledge of the environment, of the atmosphere, the land, and the seas; of things like the weather—winds, rain, snow, and particularly violent storms, hurricanes, and tornadoes, and the dangers of lightning—the sorts of things that every rural dweller quickly learns but that are foreign to most urbanites. And what of earthquakes, volcanoes, tidal waves, and other such natural causes of mass destruction? How can these be ignored in the education of all citizens? And can one really feel at ease with the environment without knowing the most elementary facts of our solar system—of planetary motions, of night and day, of the phases of the moon and the causes of eclipses and the seasons?

Such observations are (or were) second nature to all of us, and in our early school years we could account for them. Yet if pressed for explanations later on as adults, most of us fail the test, not that such failure is necessarily meaningful or diminishes our quality of life. Feeling comfortable with the basics of one's *natural* environment must be prerequisite to dealing realistically and emotionally both with science and with the environment created through technology. For simplicity, and because it makes more sense to the individual without formal schooling, we should classify this kind of knowledge in the category of *natural history*. In one way or another, everyone learns these things. Not retaining the details is less important than being aware of their overall existence and possible dangers.

Then we ask which part of the enterprise, as between science and technology, is more meaningful to the average student, and here the answer clearly points to technology. This is an age of technology—not the first, to be sure, and, in the eyes of some social historians, probably not even the most significant. But in terms of its enormous influence on all developing and already developed nations, technology in the second half of the twentieth century must surely be marked as this period's major economic and political factor. If there is anything about the scientific enterprise that can be said to affect the public directly, whether or not it is fully appreciated by most people, it would have to be technology, which the general student readily sees as more useful—and easier to grasp—than science. Thus, to the extent that content must be part of a science curriculum, such content, for the general student, is best related to technology. But *any* new curriculum for such students should revolve around the question of what makes the scientific enterprise work, particularly how and why we believe what things are true about nature. In short, what makes science successful?

To reach any objective in science education one must first overcome the most popular misconceptions about the nature of science. We know that 90 percent or more of the general public is unaware that science is primarily a method of inquiry rather than a means of accumulating facts and solving routine problems. This is a major hurdle, one that most textbooks do not address adequately, if at all, leaving it to the classroom teacher to provide the necessary synthesis. Most students learn something about the so-called scientific method, but somehow come to regard it as a fixed set of rules for scientific discovery rather than more or less characteristic features of scientific investigation. They fail to appreciate that science does not mean asking random questions of nature, that the essence of scientific inquiry is to phrase one's questions, that is, to design one's experiments, in such a manner as to obtain the most convincing answers.

Students should learn to appreciate science as a human activity. The facts of nature are there for the asking, provided one makes *appropriate* inquiries. Students easily recognize that such facts are not *made* by scientists, only uncovered by them. However, assembling the observed facts into the grand conceptual schemes that help one to understand the workings of nature and open new windows on the universe is what the business of science is really about. This aspect of science is entirely a human activity; the "laws of nature" are devised by humans, not handed down as revealed truths, and are therefore subject, like any other human endeavor, to fallibility. Yet, as we have pointed out earlier, students must be brought to realize that there is a crucial difference between the practice of science and other human activities that seek to establish fundamental truths; that only in science do we have a means of testing the truth or falsity of our statements about the universe, namely, by questioning nature itself through properly designed experiments. We should want everyone, students and adults alike, to understand that this path of continual self-correction is the single most powerful feature of the scientific enterprise. Science is not a form of "black magic." It simply demands more rigorous descriptions and rational interpretations than the commonsense explanations that fit one's everyday experiences.

What sort of topics best exemplify how and why science works

—that is, truly show the nature of the enterprise? There is nothing new or unusual about such subjects, except that for the most part they are not woven into the normal science curriculum. Yes, "scientific method," science process, and something about the philosophy of science have been popular themes in many science textbooks, more so fifty years ago than today, but mainly as an extension of the science content rather than a unifying thread. Yet is it not more important that the average student understand what is *meant* by a "scientific fact" than to *know*—and then forget—an assortment of such facts?

Or, to take another example, common discourse tends to play down the value of theories, as though they are little more than guesses. Thus, the theory of evolution is attacked by the creationists, who seek a literal interpretation of the Bible, by their saying, in effect, that evolution is *only* a theory and hence can be no better than their own theory of how man (and the Earth) evolved, or than any other theory devised by man for that matter. What the creationists ignore and expect others, including students, to disregard as well is that a theory is only as good as the range of experience it is able to account for, and here creationism comes out very poorly indeed when put to the test, compared with evolution. If only the public, and especially science teachers, had a better sense of the role of theory in the practice of science, as contrasted with its speculative connotation in everyday use! Or, to carry such misunderstanding further, probably no aspect of science brings as much resentment from humanists as its claim to exclusivity in establishing the "truth" of its statements. There are other ways, they contend, of arriving at truth.

Strangely enough, this controversy, if there really is one, turns not so much on substance but on the meaning one ascribes to the concept of truth. We do not mean here merely the formal, axiomatic truths of syllogistic arguments, that is, of deductive logic and mathematics. Nor do we limit ourselves to the truth of statements *deduced* from established laws or theories that are experimentally verifiable. Instead, we are more concerned with the truth (or validity) of grand inductive statements—with the broad generalizations (laws and theories) themselves. We have seen (chapter 3) that science has evolved simple methods of testing the truth of its statements and therefore has a reasonably clear idea of what it means by truth. But what shall be meant by truth in the humanities or in the nonscience arena generally? Can a work of art or poetry be characterized as being true or false? Here the answer is much more complex, involving a broad range of testimony on qualities, values, beliefs, and judgments which, while often commanding widespread consensus, is nevertheless not subject to *objective* proof of the kind mandated by science. Whatever one's personal view on the meaning of truth, students and adults interested in the role of science in our society must be made aware of this dichotomy.

▲ A CURRICULUM GUIDE FOR SCIENTIFIC AWARENESS

To help guide students toward the kind of scientific awareness that I believe to be the appropriate objective of general education in science, we must sharply change the emphasis of conventional curricula from science content to the process of science, continually stressing technology. The following, for example, is a list of topics that fully characterize the nature of science and should be at the core of any science curriculum intended for the general student. Indeed, even the science-bound student might well be exposed to such topics early in his or her educational career. Most have been alluded to in earlier chapters and some are elaborated above by way of example. The one framed in the form of a question was suggested by the provocative title of a recent book by Nobelist Max Perutz, which poses the challenge of how civilized society will balance the risks and benefits that derive from science.[3]

- The purpose of science
- The purpose of technology
- Are science and technology necessary?
- The meaning of scientific "facts"
- The meaning of scientific "truths"
- The role of theory in science
- The role of conceptual schemes in science
- The role of experiment in science
- The role of mathematics in science
- The complementary roles of science and technology
- The history of science, especially of technology
- The cumulative nature of science
- The horizons of science; its potential and limitations
- The threat of the antiscience and science counterculture movements

- The societal impact of science and technology
- The role of statistics
- The role of risk-benefit analysis in decision-making
- The proper use of expert science advice

Note that these topics do not deal as much with the content of science as with the *nature* of the enterprise, with matters that cut across all science disciplines and most of which underlie the practice of science. This is the sort of understanding of what science is about that we should want the educated public to be aware of—to better understand the why, what, and how of the scientific endeavor. It may seem more like philosophy than science, and to some extent it is, but these are the kinds of questions with which anyone who thinks seriously about the nature of science must grapple, and they are more likely to be imprinted in the student's mind than a lot of isolated facts about nature. Moreover, it is a lack of awareness of just such issues that leads to many of the misconceptions that people have about science and technology.

Every science curriculum, regardless of its professed goals, should at least make clear to students what science is and how it is practiced. Dispelling misconceptions about the nature of science is the first step toward true science awareness. Then follows an understanding of the nature of the enterprise. We pointed out earlier the benefit to both science and society of having the latter understand how science is practiced. There is an additional benefit to having such topics play prominent roles in the curriculum; most offer clear opportunities for student opinion and discussion, ingredients that we know make for a more gratifying classroom experience and encourage greater interest in the subject. Presently, about the only subject area that invites such student participation in conventional science courses is the STS component, if it is incorporated at all, and this, we have seen, leans more heavily on social studies than on science.

Obviously one cannot deal effectively with the nature of science solely in the abstract; it must be placed in the context of science itself, both for example and emphasis. We know that given a choice between stressing science or technology for the general student, the better choice is technology. But this poses a problem in respect to the topics shown here, for technology is not the best exemplar of many of these, while science is. It is easy to focus a curriculum on technology alone, but such a program would not convey an awareness of how science works, which should be our main objective. Hence both science and technology must have their own roles in the proposed curriculum, with the former used mainly to depict science.process, but with the actual content leaning heavily on technology.

This is more easily said than done, for it brings us full circle to the question of how to present science to the general student in a meaningful fashion, something we have not managed to do well in the past. The obvious answer, we believe, is to begin with technology, with problems that evoke familiar images of one's common experiences, and use these to work back to the underlying science needed to deal with such questions as scientific truth, laws, and theories. Technology is particularly well suited for this purpose. For example, familiar queuing problems are often used to introduce the concepts of randomness and probability theory, and problems of elevator design serve to develop the notion of acceleration and its physiological effects. Transportation systems, weather systems, food and water supplies, and personal health and safety are all of practical value (and interest) to everyone and should be part of such a curriculum, as should be the subjects of population growth and preservation of the biosphere. One could go on almost endlessly in this vein; there are very few topics in science, at least for the general, nonscience student, that cannot be introduced through familiar technology.

So much for the means of introducing science to the curriculum; what should the science consist of? Remember that our primary objective is to develop at least an awareness, if not a sound understanding, of the *nature* of science, and particularly of the topics listed above. These topics are proposed as the core of the curriculum, not as an adjunct to it. Hence careful attention must be paid to the choice of factual information and the few laws, theories, experiments, and so forth that may be used to elaborate the topics. I mentioned earlier Conant's use of examples taken from the history of science to attract student interest in the subject, on the theory that seen in retrospect the science appears much simpler.[4] There is much to be gained by this approach in the present instance, particularly as some of the basic laws and theories that hold today actually predate the Industrial Revolution, and many of the important ones were developed before the start of the twentieth century. Conant's focus was on science as history. More interesting (and meaningful) to students, however, might have been selected topics in the history of technology, with special emphasis on how these technologies grew out of scientific discoveries of the time.

In summary, then, the proposed procedure is as follows: (a) introduce *all* topics through some relevant problems or issues in technology, but only where these are meaningful to students; (b) work back to the underlying science where, and only to the extent, it is needed to account for the technology; (c) use the underlying science, where appropriate, as a springboard to discuss the nature of the scientific enterprise—namely, the role of experiment and the meaning of scientific truth, facts, laws, theories, etc.; (d) return to technology, again where appropriate, as the basis for discussing the science/society interface; and finally (e) conclude with when and how to use expert advice at the science/society interface. It is not my purpose to write the textbook here, but I hope that when such a curriculum is finally implemented, it will continually stress the *nature of science* as its major theme, all the while focusing on technology, and that it will include as many of the topics listed above as is practicable.

The final, and in some ways the most important question, is the depth in science to which one should try to carry the general student. Here, based on past experience, we know that there are some practical limits. We know, for example, that quantitative analysis must be kept to a minimum, that beyond seeking to demonstrate the vital role of mathematics in the practice of science we cannot expect students to be proficient in using it, at least not until major changes occur in the way that mathematics is taught—if then. Nor can we expect critical thinking to be practiced at all times by the average student, but we should expect teachers to be endowed with this mental skill. Hence the pure science portion of the curriculum, as we normally understand its meaning, should be kept to a bare minimum, consistent with the requirements that (a) it helps to illuminate the nature of the enterprise, and (b) it is meaningful to the student. For example, in dealing with the science/society interface it is clearly unnecessary to go into great depth about the science or technology involved.

We can all agree that education designed to increase awareness of global environmental problems should have a top priority. But is it necessary to understand the science of acid rain or the greenhouse effect to have such awareness? Is it not sufficient to know their ultimate causes, their solutions, and the economic and political changes likely to result from these solutions? Environmental problems are almost never politically neutral; one cannot wish them away or even easily legislate them away, the neo-Luddites notwithstanding, without taking into account their economic and political implications. How much of a lower living standard, for example, would free nations be willing to accept to be rid of environmental pollution? The United States currently consumes about 28 percent of the world's oil production, yet has only 2 percent of its population and about 6 percent of the land mass. Would Americans be willing to get along on only 2 percent, or even 6 percent, of the oil supply without having equivalent energy sources to replace it? The economic and political implications of such demographics are clearly part of the awareness that society needs, but an *in-depth* understanding of the science or technology involved in these environmental problems is unlikely to contribute much to the steps that society can or should take to resolve them.

A further word is in order here on the STS movement, which many will recognize as containing the same elements as the curriculum outline proposed above, namely, science, technology, and society. However, there is a subtle but very important difference between the two. As we have already seen, STS starts at the technology/society interface, that is, with social or political issues that provoke some segment of society and that have a scientific or, more often, a technological base. This is the primary rationale of STS—an introduction to science through its effects on society. Since the effects of technology that cause society the most concern are considered deleterious, the STS approach most often turns into a critique of science and technology, and of their practitioners. While its conceptual framework has a certain appeal, and the technology/society interface should be prominent in any modern science curriculum for the general student, in STS science and technology are subordinated to the societal component, which dominates the curriculum, gives it a strong social studies flavor, and must eventually lead to its downfall as a surrogate for science education. Bear in mind that many of the important concepts and developments in science and technology do not present a societal problem that warrants their inclusion in a typical STS curriculum. Hence, one needs a far reach to develop a well-rounded science program based on the STS philosophy.

In sharp contrast, the curriculum we propose begins and ends with technology. Its intent is to help students reach a comfortable compact with the everyday technology that surrounds them, and with enough of its underlying science to appreciate how the overall enterprise works. Only then is the student prepared to deal rationally and intelligently with the societal issues stemming from science or technology.

▲ SOME CAVEATS

Every science curriculum has its problems, and my proposed guide is no exception. To be successful the guide presupposes that (a) the students have a reasonable command of the traditional 3 R's, (b) they are at least passively receptive to science, and (c) their teachers have a "worldview" of science and technology rather than of a narrowly defined science discipline. In some respects the last is the most important prerequisite, for to handle this or any good science curriculum for the general student properly, we need teachers who not only understand the science and in this case the technology as well, but who also appreciate their cultural, humanistic, and societal aspects. And if the teacher is a specialist in one branch of science, it is essential that he or she is able to see across disciplinary borders. This is a very tall order indeed, and few teachers are able to fill it at any of the precollege levels, particularly in the elementary and middle schools. But even at the high school level, where the science teacher is normally expected to be better prepared in the discipline than at the lower grade levels, the variability among science teachers throughout the country is appalling and inexcusable. Nor is the college level, where so many science faculty regard teaching nonmajors as little more than a necessary chore, exempt from such criticism. We must emphasize again the central role of the teacher rather than the curriculum in quality science education. To a highly qualified teacher the curriculum is really no more than a floor plan; to most teachers, however, it is an absolute necessity, regrettable as that may be. Thus, the curriculum remains an essential feature of science education today.

Epilogue

> Alas! Hegel was right when he said that we learn from history that men never learn anything from history.
> —G. B. Shaw (*Heartbreak House*, 1917)

▲ The main thrust of this book has been to show that the common conception of scientific literacy as a goal of general education in science cannot possibly be realized, and that if science education is to have any real meaning for this large group of students, its objectives must be sharply redefined and modified. If we take scientific literacy to mean (a) having an awareness of how the science/technology enterprise works, (b) having the public feel comfortable with knowing what science is *about*, even though it may not know much *about* science, (c) having the public understand what can be expected from science, and (d) knowing how public opinion can best be heard in respect to the enterprise, we will have defined a set of objectives that actually may be attainable. This form of functional literacy is out of step with virtually all current efforts to achieve scientific literacy, as these efforts continue to focus on traditional science knowledge as the mark of a literate individual. However, it is one that may turn out to be far more useful to everyone concerned—the public as well as the science and science education communities—than all our previous efforts to bring the general public into the mainstream of science and technology. Rather than leaving technology as an offshoot of science, our answer is to *focus* instead on technology as the starting point of a science curriculum for the general student.

This is by no means an original thought. Technology education has been tried several times, but without much success; the main reason, one suspects, is that it always had to compete with the ever-elusive goal of "scientific literacy." In any event, technology has thus far failed to catch on, especially with science educators, as a basic curriculum for the general student. But it has not been fairly tested. Where the normal science curriculum includes too little technology, the typical technology curriculum contains too little about the nature of science. Both are essential, but the mix must be carefully drawn from the topics I proposed earlier to form a cohesive picture of the overall scientific enterprise.

▲ NO SIMPLE SOLUTIONS

It should be clear that there are no *easy* answers to the problem, not even the one proposed here, which we have seen carries its own burden of uncertainties. The more optimistic educators, while possibly conceding that science education for the general student has thus far missed its mark, continue to believe that it is only a matter of finding the right formula, that continued experimentation with new curricula, new modes of instruction, and some of the new educational technology will eventually result in truly effective general education in science. Yet past experience strongly suggests that a progression of incremental curriculum "improvements," apparently favored by most as the surest and least risky path toward an ultimate literacy in science for students and adults, cannot achieve the desired result. One reason is that the "lag phase" for each incremental change is so long. Years are usually required to introduce, refine, and *properly* evaluate·every new science segment or course, let alone an entire curriculum. And if one should conclude, after such testing, that a new course is indeed an improvement over the old, think how often this process must be repeated, each time with a *successful* outcome, before the cumulative result becomes truly noticeable. And before this can occur another formidable barrier often stands in the way; namely, new generations of students and educators appear on the scene, ready to discard the old in favor of still another round of experimental programs. Moreover, all this must be done while fighting an uphill battle against the obvious lack of incentive on the part of most students to absorb the kind of "literacy" we keep thrusting upon them.

The more pragmatic educators, we have seen, realize that curriculum reform alone, without proper motivation of students and teachers, is not the solution in science education. And at the extreme end of the science education spectrum are the most pessimistic educators who fear that, barring a major shift in socioeducational mores, science as a *compulsory* subject in our high schools may eventually go the way of

Latin as a required language. That is, unless science becomes widely accepted as an educational imperative rather than simply being tolerated, both for high school graduation and college admission, its luster will gradually tarnish and ultimately it will become an elective subject for all but science-bound students. It has survived till now not because it captures the minds of many students, which it obviously fails to do, but partly because of the fantasy and idealism surrounding the notion of scientific literacy for the masses, and mainly because of the importance of science to highly industrialized nations. And it will continue to grow because of this need. However, failing some extraordinary change in the avowed purpose of science education for the nonscience student, the general population is likely to become even more of a bystander than an active participant in science-based matters.

The point has been made that the only truly essential and enduring feature of science education is our continuing need to produce new scientists and engineers—to educate the science-bound students. Most science educators feel that we know much more today about the problems of teaching science and how to improve science curricula than our predecessors did even a half century ago, yet we cannot claim to be any nearer the goal than they were. We certainly know more about child development, learning theory, cognitive science, teacher training, and evaluation—indeed, all the tools that should have helped solve the puzzle by now—except one: we have been trying to solve the wrong puzzle! The science education community keeps insisting that the public conform to the *traditional* meaning of scientific literacy, that is, to knowing some "textbook" science instead of sharply redefining such literacy to conform with the true public need.

I contend that a sharp change in direction will be needed to budge science education from its present tradition-ordained path, and what I have proposed is just such a departure, both in the objectives of general education in science and in the manner of achieving that goal. The curriculum guide suggested in the previous chapter is only a rough road map; the real challenge lies in putting the pieces together into individual courses and a full-scale curriculum for the general student. There is a caveat, however. We know that curriculum changes alone have not solved the problem in the past—the wrong problem, it turns out, but an educational problem nonetheless. It is remarkable how firmly educators cling to the notion that the focal point of an academic subject is the curriculum, far outweighing all other factors. In a sense this may be true, but only to the extent that the curriculum defines the subject matter of a course of study and provides stepping-stones to the ultimate goal. It serves as the starting point, but is rarely the end point as well; it might be considered a necessary but not a sufficient condition for effective education. Hence it is entirely possible that the new curriculum approach proposed here, barring other needed changes in our educational system such as better teaching, improved facilities, and greater support for education generally, may be equally ineffectual. However, I believe that on balance it offers a better chance of succeeding because it addresses the real problem, namely, setting a *meaningful* goal for "scientific literacy," one that may capture the imagination and interest of all parties in the venture: students, teachers, and the general public.

▲ THE ROLE OF THE FEDERAL GOVERNMENT

As we move toward a fresh look at the question of science education for the masses, we should bear in mind a number of peripheral issues that are likely to affect education generally (as well as science education) in the next decade. Probably the most important of these is the role that will be played by the federal government. Responding to the widespread complaints over the state of precollege education in America, former president George Bush, we recall, announced that he wished to be known as the "Education President." He accordingly initiated several steps that, if followed through with the good of education uppermost in mind rather than politics, offered the potential of triggering a meaningful reform movement in U.S. education.[1] His proposed timetable, which was wholly unrealistic, was unachievable, particularly his goal to make U.S. students the best in the world in science and mathematics by the year 2000.

True educational reform does not move that fast, especially in the United States, with its fifty separate school systems instead of a single one, which, in principle at least, should be easier to manage. The only effective role of the federal government, mainly through its National Science Foundation and the Department of Education, is to provide educational surveys, financial aid to schools, and grants to underwrite new experimental programs in education. While playing an important role in developing innovative science curricula, these agencies lack the legal authority to force their adoption by local school systems, which are subject only to the directives of their state education departments. Where federal guidelines or national curriculum developments are adopted by local schools, they do so only on a voluntary basis. However, there is a powerful, though indirect, method that the federal government can use to impose its educational objectives on the states, namely, through its funding capacity.

Not to be outdone in the education area, the Clinton administration was quick to announce its "Goals 2000" program, which called for setting high national standards for student performance in all basic areas, including math, reading, writing, and science. As far as the standards of science education are concerned, the movement is under the direction of the National Academy of Sciences' Committee on Science Education Standards and Assessment, with the cooperation of various professional science teacher groups and educators. To be considered truly "high standards" in science, one must assume that these will engage students in more ways than the usual factual recall and solution of standard problem types, and critical thinking will have to be prominent among the evaluation criteria. To encourage states to raise their own standards to conform with the national levels, federal grants would be provided. However, to put some teeth into the plan, presumably such grants would be awarded to a state only as long as it can demonstrate through objective testing that the performance of its students is improving.

The question, of course, is whether the standards that are finally agreed upon will be meaningful or so watered down as to be ineffectual. The weakest link in the chain is not so much the standards themselves but the testing methods used to determine whether the standards are being met by a given state or local school district. Here, as always, one can expect to find strong differences among evaluation experts on the interpretation of performance-based testing, as well as among parents and special interest groups who object to such tests as an infringement of their personal views or beliefs, generally more so in the humanities and social sciences than in natural science. A major reason is that new tests, such as some in critical thinking and rational behavior, are being devised to measure factors that formerly were not included in the usual evaluation of student performance. This disturbs "back-to-basics" advocates and many conservatives who believe that education has already strayed too far from its proper goals. As of this writing, early in the process of developing the standards program, the controversy is already in full swing, and promises to intensify before playing itself out.

That some form of national standards will be adopted, at least in math and science, is almost a foregone conclusion, given that the government is prepared to back up its goals with federal grants. Whether these will have the desired effect of upgrading science education on the state level is not as certain, given the autonomy of state education departments and the differences that are likely to evolve during the extensive review process. In any event, even assuming that the compromise standards turn out to be meaningful, what bearing might they have on the central problem of scientific literacy, other factors remaining unchanged? Performance standards alone cannot be expected to turn otherwise uninterested students into scholars. However, their indirect effect on improving the science understanding of teachers might be very significant, which should help to improve science instruction generally—and possibly the acceptance of science by students. On this score alone the idea warrants the support of the science education community.

▲ PRIVATIZING THE SCHOOLS?

One way of improving education may lie in privatizing our public school system to achieve greater efficiency through the free play of the marketplace. Unpopular as this notion may be to most educators, the fact is that most private schools in the United States, both elementary and secondary, while not run as strictly commercial enterprises and generally not profit-oriented, nevertheless provide high-level (some would say the very best) education to a relatively small fraction (about 12 percent) of our total school population. They manage to do this at a per pupil expenditure averaging only about 70 percent that of our public schools.[2] Some would say that this is because teachers are paid less in private schools, which is often true but does not tell the full story.

Teachers in private schools seem to have a greater sense of professionalism and personal satisfaction in their work than their counterparts in the public schools, despite lower salaries in many cases. And at the college level, where 22 percent of all students are enrolled in private schools, there is no question about the quality of most such institutions, or of the educational experience they provide. Indeed, there is little difference between most well-funded public and private four-year colleges in terms of the basic quality of their faculties and facilities.

Of course, the private schools, because they are able to select their students, do not face the varied problems of public education, such as handicapped (physically, mentally, or emotionally, who comprise roughly 10 percent of all public school students) or marginal students, bilingual or other special education students, busing at public expense, disciplinary problems, and so forth, yet one is tempted to believe that given the opportunity, the private sector might do a better job of educating our children, at the same or lower cost than the current system of public education manages to do under community and state control. By this I do not mean private, for-profit ventures competing with the public schools and with one another, but private, either for-profit or not-for-profit management contracts awarded by municipalities to companies approved and accredited by appropriate state agencies whose performance is monitored by the state but who, in the final analysis, are answerable to the community. There could be some real advantages to this. Not only might the overall costs be lower, but if widely adopted, the freedom to share specialists across district lines among neighboring communities could provide a significant educational advantage. It poses many problems, of course, but none that appear, at least on the surface, to be insurmountable. Some movement in this direction is already under way, with education in several communities contracted to private management groups. While reactions to date are mixed, not solely on the question of cost but on the quality of education provided by the private companies, one can project that the privatization movement will continue to grow until evidence of its worth becomes much clearer than it is today.

A possibly interesting variant of privatization in the public schools might be in science education, where it has proved particularly difficult to staff all schools with highly qualified teachers, and laboratory facilities are expensive and often underutilized. Here, one can imagine that private management on a communitywide basis should be able to provide trained science specialists and better equipment and supplies via a mobile system, at least to all the elementary and middle schools in the community, schools that typically lack adequate science teaching and science facilities.

▲ EQUITY IN SCIENCE AND ENGINEERING

Since its beginnings, science in the United States, as in all industrialized nations, has been male-dominated, particularly in the physical sciences and engineering. Even in the post–World War II period, when scientists and engineers were in short supply, parents still advised their daughters to avoid science, and especially engineering, as a career objective, advice that was generally echoed by high school guidance counselors. This gender imbalance was so pervasive and so obvious to everyone in education, even as barriers were being broken down in most fields, that thirty years ago Alice Rossi was prompted to ask: "Women in science, why so few?"[3]

The situation has improved somewhat in recent years, propelled partly by more stringent sex discrimination legislation and equal opportunity activism, and partly by funding programs designed to attract women to these fields. But we are still very far from gender equality in science and engineering. The inequity varies widely by field, being much less pronounced in the life sciences and mathematics, but positively shocking in physics and engineering. For instance, by 1985 the fraction of bachelor's degrees earned by women in the life sciences had reached close to half the total in those fields, the same as the ratio of all bachelor's degrees earned by women that year; this ratio in the life sciences had grown from 30 percent just fifteen years earlier. Similarly, in mathematics the ratio was about 46 percent. Even in chemistry, and also in computer science, the ratio was about 37 percent. But in both physics and engineering only 15 percent of the degrees were awarded to women.[4]

Contrast this with the health professions, which are dominated by women. Fully 85 percent of the bachelor's degrees awarded in this area in 1985 went to women (mostly in nursing), just the reverse of their presence in physics or engineering. One of the reasons often cited for this huge disparity is the desire of women to avoid disciplines that rely heavily on mathematics. There may be some truth to this, but it fails to account for the substantial number of women who major in mathematics—very nearly the same as the number of men.

It is more likely that the reasons women tend to avoid careers in physics and engineering have to do more with self-image and with their perceptions of the types of careers offered in these fields than with an unwillingness to cope with the mathematical rigor involved. If the career opportunities were truly attractive to women, as they may become one day, especially if a shortage of professionals should occur in these fields, we should witness many more women going into the physical sciences and engineering.

▲ LEARNING FROM THE PAST

We are now nearing the end of the twentieth century, one marked by an enormous growth in scientific knowledge and almost two centuries of debate, largely by British scholars, on the relative merits of science versus the humanities as the focal point of our educational system. And as the breadth and depth of science increased so dramatically, so too did the drive to have it better understood by the common citizen—unmindful of the fact that the problems of doing so multiplied with each new scientific discovery. Ever since science became part of the general school curriculum in America early in this century, the lofty theme of "science for the masses" has been repeated time and again, and in many guises, by prominent educators, the latest being the concerted call for *scientific literacy*. Unfortunately, almost from the very beginning, the two threads of science education, the "why" and the "what," became intertwined and have remained so to this day. *Why* should science be required of all students, and *what* should it consist of? The "why" was pretty much taken for granted and all the emphasis was placed on curriculum design.

In the past century alone, we have moved from instruction in practical science through science instruction as a means of developing sound mental habits, termed by John Dewey, the famous early twentieth-century educational philosopher, "scientific habits of the mind," through science instruction as a cultural imperative, meaning broadly the process of science as seen through the eyes of scientists, through technology as an entrée to scientific knowledge, and finally to science instruction as a societal imperative, usually labeled "scientific literacy" or "public understanding of science" and generally meaning an understanding of science adequate for the populace to deal with related social issues. Along the way we have found all manner of variations and combinations. These include the new British experiment that sets science on an academic pedestal, and the massive curriculum reform programs in the United States that have been undertaken by, among others, the National Science Teachers Association, with its Scope, Sequence and Coordination (SS&C) Project, and the American Association for the Advancement of Science, with Project 2061. Throughout most of this activity, however, the "why" has remained submerged. Science education for all is taken as a given, the only question being, What science should be taught? Regrettably, the one solution that might have proved effective, namely technology education, failed to receive the full support it needed from the scientific community, which has long tended to regard technology studies as intellectually inferior to pure science.

Earlier, we discussed the views of Spencer, Huxley, Poincaré, Dewey, Bronowski, and Snow, presented as but a few of the many distinguished scientists and educators who have tried to persuade society that, for one reason or another, *everyone* should be versed in science to some degree. Yet we seem to have learned very little from the dismal history of these efforts, for the science education community still clings to the romantic notion that everyone, not merely the 5 to 10 percent of our youth who choose careers in science or engineering, should be required to study science. To make matters worse, we keep insisting that public understanding of science means understanding some *basic science* rather than the technology that the public finds more palatable. All this despite the fact that ever since the Enlightenment some 250 years ago, society has been sending back the same message: give us the useful end products of science, as long as they cause us no real harm; but while we can relate to their technology, don't *require* that we understand their underlying science. Despite all the evidence that the general public seriously questions the value of the science education we have been forcing upon it in our schools, instead of responding to its perceived needs, we insist upon presenting our old-fashioned notions of science understanding, with minor variations, to an unwilling audience. Surely, it is time we take this message to heart, and, before we spend many billions of dollars more on loosely targeted school science programs, we should seriously question both the *why* and the *what* of compulsory science education.

▲ NOTES

▲ CHAPTER 1. *A Crisis in Science Education?*

1. National Science Foundation, *Today's Problems, Tomorrow's Crises*, Report of the National Science Board Commission on Precollege Education in Mathematics, Science and Technology (Washington, D.C.: U.S. Government Printing Office, October 1982).
2. F. Hechinger, *New York Times*, October 4, 1983.
3. U.S. Department of Education, *A Nation at Risk: The Imperative for Educational Reform*, Report of the National Commission on Excellence in Education (Washington, D.C.: U.S. Government Printing Office, April 1983).
4. A. Champagne, *Proceedings of NSTA/ASE Joint Seminar on Science Education* (York, U.K., 1986), 240–256.
5. M. H. Shamos, "A False Alarm in Science Education," *Issues in Science and Technology* 4 (3) (1988): 65–69.
6. U.S. Department of Education, National Center for Educational Statistics, *Condition of Education* (Washington, D.C.: U.S. Government Printing Office, 1985).
7. U.S. Department of Education, *The Science Report Card: Elements of Risk of Recovery*, National Assessment of Educational Progress (Washington, D.C.: U.S. Government Printing Office, 1988).
8. National Science Foundation, *Science Achievement in Seventeen Countries: A Preliminary Report* (Washington, D.C.: U.S. Government Printing Office, 1988).
9. C. F. Citro and G. Kalton, eds., *Surveying the Nation's Scientists and Engineers—A Data System for the 1990s*, Panel to Study the NSF Scientific and Technical Personnel Data of the Committee on National Statistics, National Research Council (Washington, D.C.: National Academy Press, 1989).

239

10. R. White, presidential address, National Academy of Engineering, October 2, 1990.
11. A. Fechter, "Engineering Shortages and Shortfalls: Myths and Realities," *The Bridge* 20:2 (1990): 16.
12. R. Atkinson, "Supply and Demand for Scientists and Engineers: A National Crisis in the Making," speech delivered at annual meeting of the AAAS, New Orleans, 1990.
13. National Science Foundation, *The Science and Engineering Pipeline*, PRA Report 87–2 (Washington, D.C.: U.S. Government Printing Office, April 1987).
14. R. B. Reich, "Metamorphosis of the American Worker," *Business Month*, November 1990, 58–66.
15. The Age Discrimination in Employment Act, as amended in 1986, provides a special exemption for institutions of higher education.
16. See, e.g., M.L.A. MacVicar, "Biting the Bullet on Science Education," *Issues in Science and Technology* 7, no. 1 (1990): 36. Note that new legislation that eases immigration for aliens with advanced degrees or "exceptional ability" in, among other fields, the sciences has been adopted.
17. P. E. Stephan, at AAAS Annual Meeting, Washington, D.C., February 1991.
18. S. G. Levin and P. E. Stephan, "Research Productivity Over the Life Cycle: Evidence for American Scientists," *American Economic Review* 81 (1991): 114–132.

▲ **CHAPTER 2.** *Public Understanding of Science*

1. Plato, *The Republic*.
2. Jacob Bronowski, *The Common Sense of Science* (Cambridge, Mass.: Harvard University Press, 1958), 109.
3. See W. Dampier, *A History of Science*, 4th ed. (Cambridge, U.K.: Cambridge University Press, 1948), 105.
4. Cicero, *De Oratore*, Book I.
5. L. P. Williams, "Science, Education and Napoleon I." *Isis* 47:4 (1956): 369–382.
6. Galileo Galilei, *Dialogue Concerning the Two Chief Systems of the World, the Ptolemaic and the Copernican* (1630) (see English trans. by T. Salusbury, 1661).
7. F. Bacon, *Novum Organum*, ed. J. Devey (New York: P. F. Collier and Son, 1901).
8. Williams, "Science, Education and Napoleon I," 369–382.
9. T. H. Huxley, "Science and Culture," in *Science and Education* (New York: Philosophical Library, 1964), 120.
10. H. Spencer, *Essays on Education, Etc.* (London: J. M. Dent and Sons, 1911), 3–44.

11. H. Spencer, *Autobiography*, vol. 2 (New York: D. Appleton, 1904), 42–44.
12. T. H. Huxley, "A Liberal Education; and Where to Find It," in *Science and Education* (New York: Philosophical Library, 1964), 72.
13. Huxley, "Science and Culture," 126–127.
14. M. Arnold, *Discourses in America* (London: Macmillan, 1885).
15. J.B.S. Haldane, *Daedalus; or Science and the Future* (New York: E. P. Dutton, 1924).
16. H. Poincaré, *The Value of Science*, trans. George B. Halstead (New York: Science Press, 1907), 8.
17. B. Franklin, *Writings*, ed. A. H. Smyth, vol. 2 (New York: Macmillan, 1907), 395.
18. J. S. Roucek, *The Challenge of Science Education* (Freeport, N.Y.: Books for Libraries Press, 1959), 85.
19. Franklin, *Writings*, 395.
20. A. de Tocqueville, *Democracy in America*, trans. Henry Reeve (London: Saunders and Otley, 1838).
21. J. C. Burnham, *How Superstition Won and Science Lost* (New Brunswick, N.J.: Rutgers University Press, 1987), 129.
22. J. C. Greene, *American Science in the Age of Jefferson* (Ames: Iowa State University Press, 1984).
23. J. Lambert, *Scientific and Technical Journals* (London: Clive Bingley, 1985).
24. Burnham, *Superstition*, 136.
25. W. D. Miles, "Public Lectures on Chemistry in the United States," *Ambix* 15 (1968): 141.
26. G. H. Daniels, *Science in American Society* (New York: Alfred A. Knopf, 1971), 265.
27. Commission on Secondary School Curriculum, *Science in General Education* (New York: Appleton-Century, 1938), 8.
28. See, e.g., F. W. Clarke, "A Report on the Teaching of Chemistry and Physics in the United States," U.S. Bureau of Education, Information Circular No. 6 (1880), 10; "Notes to Teachers," Report of the Committee of Ten on Secondary School Studies, vii (Published for the National Education Association by American Book Company, 1894).
29. M. B. Zuckerman, "The Illiteracy Epidemic," *U.S. News & World Report*, June 12, 1989, 72.

▲ **CHAPTER 3.** *The Nature of Science*

1. N. Campbell, *What Is Science* (New York: Dover Publications, 1921).
2. G. G. Simpson, "Biology and the Nature of Science," *Science* 139 (1963): 81.
3. E. Nagel, *The Structure of Science: Problems in the Logic of Scientific Explanation* (New York: Harcourt, Brace and World, 1961).

4. See, e.g., B. Russell, *An Inquiry into Meaning and Truth* (New York: W. W. Norton, 1940), 345.

5. Much like that described by Democritus in the fifth century B.C.: "In truth, there are atoms and there is a void."

6. Indeed, this is what is meant by a "sharp" knife or a "sharp" point: one of such small dimension that it cannot support much negative charge (electrons).

7. A similar example was used by Eddington in the first of his series of Gifford lectures at the University of Edinburgh, Scotland; see A. S. Eddington, *The Nature of the Physical World* (London: Macmillan, 1928).

8. Nagel, *Structure of Science*, 98.

9. Bronowski, *Common Sense of Science*, 148.

10. There is no clear-cut distinction between these two terms except that laws are generally reserved for experimentally determined relationships while "theoretical laws," or simply theories, are applied to relationships resulting from assumptions (conceptual schemes) devised to account for a given phenomenon. For example, the well-known gas *laws* may be determined by measuring the pressure, volume, and temperature of a gas under varying conditions, while the kinetic *theory* of gases, which assumes a gas to be made up of moving molecules that exert a force on the walls of its container, yields a theoretical derivation of the same gas laws.

11. See, e.g., A. Lightman and O. Gingerich, "When Do Anomalies Begin?" *Science* 255 (1991): 690–695.

12. J. B. Conant, *Science and Common Sense* (New Haven: Yale University Press, 1951).

13. L. K. Nash, *The Nature of the Natural Sciences* (Boston: Little, Brown, 1963), 83.

14. K. Popper, *The Logic of Scientific Discovery* (New York: Basic Books, 1959; trans. from the original, which was published in 1934).

15. L. Cranberg, "What About ESP?" Letter to Editor, *American Journal of Physics* 59, no. 2 (1991): 104.

16. J. Priestley (1733–1804), *The History and Present State of Electricity* (1767).

17. F. Bacon, *Novum Organum* (1620).

18. K. Pearson, *The Grammar of Science* (New York: Meridian Books, 1957; published initially in 1892, revised in 1900, and final edition published in 1911).

19. M. Born, *Experiment and Theory in Physics* (London: Cambridge University Press, 1934), 44.

20. T. H. Huxley, "We Are All Scientists," in *Darwiniana* (New York: Appleton-Century, 1863).

21. D. Hume, *Works*, chap. 4, 439–441. Quoted from A. Wolf, *History of Science*, 2d ed. (London: Allen and Unwin, 1952), 761.

22. Nagel, *Structure of Science*.

23. D. Hofstadter, CSICOP Conference, Chicago, November 5, 1988. See *The Skeptical Inquirer* 13 (Spring 1989): 3.

24. T. H. Huxley, "The Method of Zadig," in *Collected Essays*, vol. 4 (New York: Appleton and Co., 1896).

25. D. J. Boorstin, "The Shadowland of Democracy," *U.S. News & World Report*, November 14, 1988, 61.

26. Poincaré, *Value of Science*, 8.

27. William of Ockham (or Occam), 1300?–1349?, an English scholastic philosopher who formulated the doctrine that "entities must not be unnecessarily multiplied."

28. See, e.g., W. M. Thorburn, "The Myth of Occam's Razor," *Mind* 27 (1918): 345–353; W. H. Jefferys and J. O. Berger, "Ockham's Razor and Bayesian Analysis," *American Scientist* 80 (1992): 64–72.

29. Quoted from the *Catalogue of Portsmouth Papers* (Cambridge, U.K.: Cambridge University Press, 1988), xviii. The reason he found the two to agree only "pretty nearly" can be attributed to the fact that he used an inaccurate estimate of the size of the earth.

30. Poincaré, *Value of Science*, 84.

31. M. H. Shamos, *Great Experiments in Physics* (New York: Holt, Rinehart and Winston, 1959).

32. Karl Pearson, David Hume, and John Locke being noteworthy in this connection.

33. The "Uncertainty Principle," which shows that on this scale the act of measurement perturbs the observation to an extent that limits the precision of the measurement.

34. P. W. Bridgman, *The Nature of Physical Theory* (Princeton, N.J.: Princeton University Press, 1936).

35. N. Machiavelli, from *The Prince* (1513), chap. 21.

▲ **CHAPTER 4.** *The Scientific Literacy Movement*

1. Harris Public Opinion Poll, U.S. Office of Technology Assessment, Washington, D.C. (1986).

2. D. Roy et al., eds., *Bioscience—Society*, Schering Foundation Workshop (Chichester, U.K.: John Wiley and Sons, 1991), 381.

3. D. J. Boorstin, "Remarks on Receiving ΦBK's Distinguished Service to Humanities Award," *The Key Reporter*, Winter 1988–89, 6.

4. P. J. Kuznick, *Scientists as Political Activists in 1930s America* (Chicago: University of Chicago Press, 1987).

5. D. S. Greenberg, *The Politics of Pure Science* (New York: New American Library, 1967), 125.

6. J. R. Oppenheimer, lecture at M.I.T., Cambridge, Mass., 1947.

7. J. Dewey, "Symposium on the Purpose and Organization of Physics

Teaching in Secondary Schools, Part 13," *School Science and Mathematics* 9 (1909): 291–292.

8. Dewey, "Symposium."

9. K. Pearson, *The Grammar of Science*, 2d ed. (London: Adam and Charles Black, 1900).

10. A. B. Champagne and L. E. Klopfer, "A Sixty-Year Perspective on Three Issues in Science Education," *Science Education* 61, no. 4 (1977): 431–452.

11. J. Dewey, "The Supreme Intellectual Obligation," *Science Education* 18 (1934): 1–4.

12. J. Dewey, *How We Think: A Restatement of the Relation of Reflective Thinking to the Educative Process* (Boston: D.C. Heath, 1933). Note that Dewey's "reflective thinking" should not be confused with Kant's "reflective judgment." The latter is more closely related to reasoning by induction, while Dewey's term was intended to encompass more of the process of science, or "scientific method."

13. J. D. Miller, "Scientific Literacy," speech delivered at annual meeting of the AAAS, San Francisco, January 1989.

14. "A Wisconsin Philosophy of Science Teaching," *Wisconsin Journal of Education*, November 1932.

15. I. C. Davis, "The Measurement of Scientific Attitudes," *Science Education* 19 (1935): 117–122.

16. Davis, "Scientific Attitudes."

17. A.S.G. Hoff, "A Test for Scientific Attitude," *School Science and Mathematics* 36 (1936): 763–770.

18. V. H. Noll, "Measuring the Scientific Attitude," *Journal of Abnormal and Social Psychology* 30 (1935): 145–154.

19. J. C. Burnham, *How Superstition Won and Science Lost* (New Brunswick, N.J.: Rutgers University Press, 1987).

20. B.S.P. Shen, "Scientific Literacy and the Public Understanding of Science," in *Communication of Scientific Information*, ed. S. Day (Basel: Karger, 1975).

21. K. Prewitt, "Scientific Illiteracy and Democratic Theory," *Daedalus* 112, no. 2 (1983): 49–64.

22. J. B. Conant, foreword in I. B. Cohen and F. G. Watson, eds., *General Education in Science* (Cambridge, Mass.: Harvard University Press, 1952).

23. Miller, "Scientific Literacy."

24. M. H. Shamos, "Views of Scientific Literacy in Elementary School Science Programs: Past, Present, and Future," in *Scientific Literacy*, ed. A. Champagne et al. (Washington, D.C.: AAAS, 1989), 109–127.

25. E. D. Hirsch, *Cultural Literacy: What Every American Needs to Know* (Boston: Houghton Mifflin, 1987).

26. E. D. Hirsch, J. F. Kett, and J. Trefil, *The Dictionary of Cultural Literacy* (Boston: Houghton Mifflin, 1988).

27. Note that the apparent success of the Patriot radar-guided missiles in intercepting incoming missiles during the 1991 Persian Gulf War, while still debatable as to their actual accomplishment, revived the Star Wars debate. But it was significantly modified to involve antimissile missiles, rather than powerful X-ray lasers, to destroy incoming enemy missiles.

28. M. H. Shamos, "The Lesson Every Child Need Not Learn," *The Sciences* 28, no. 4 (July/August 1988): 14–20.

29. J. A. Paulos, *Innumeracy: Mathematical Illiteracy and Its Consequences* (New York: Hill and Wang, 1988); idem, *Ruminations of a Numbers Man* (New York: Alfred A. Knopf, 1990).

30. A. Astin and H. Astin, *Undergraduate Science Education: The Impact of Different College Environments on the Educational Pipeline in the Sciences* (Los Angeles: Report of Higher Education Research Institute [HERI], 1993).

31. D. Harman, *Illiteracy: A National Dilemma* (New York: Cambridge Book Company, 1986).

32. Reich, "Metamorphosis of the American Worker," 58.

▲ **CHAPTER 5.** *The "Two Cultures"—and a Third*

1. Allan Bloom, *The Closing of the American Mind* (New York: Simon and Schuster, 1987).

2. Hirsch, *Cultural Literacy*.

3. C. P. Snow, "The Two Cultures," *New Statesman* 6 (October 1956); *The Two Cultures and the Scientific Revolution* (Cambridge, U.K.: Cambridge University Press, 1959). Given as the Rede Lecture, May 1959.

4. J. Bronowski, "The Educated Man in 1984," *Science* 123 (1956): 710–712.

5. G. Orwell, *Nineteen Eighty Four* (New York: Harcourt Brace, 1949).

6. A. Huxley, *Brave New World* (Garden City, N.Y.: Doubleday, Doran, 1932).

7. C. P. Snow, *The Two Cultures* (Cambridge, U.K.: Cambridge University Press, 1959).

8. C. P. Snow, *The Two Cultures and a Second Look* (Cambridge, U.K.: Cambridge University Press, 1963).

9. *The Novo Report: American Attitudes and Beliefs about Genetic Engineering* (New York: Novo Information Center, 1987); *New Developments in Biotechnology: Public Perceptions of Biotechnology* (Washington, D.C.: U.S. Office of Technology Assessment, 1986).

10. Huxley, *Brave New World*.

11. Spencer, *Essays on Education, Etc.*

12. Shamos, "The Lesson Every Child Need Not Learn," 14–20.

13. L. Winner, *Autonomous Technology* (Cambridge, Mass.: MIT Press, 1977), 199.

14. F. Capra, *The Tao of Physics* (New York: Bantam Books, 1977).

15. M. Ferguson, *The Aquarian Conspiracy* (Los Angeles: J. P. Tarcher, 1980).

16. The best source exposing pseudoscientific and paranormal claims is *The Skeptical Inquirer*, the official journal of the Committee for the Scientific Investigation of the Paranormal, Paul Kurtz, Chairman (Box 229, Buffalo, NY 14215-0229).

17. C. Glendinning, "Notes Toward a Neo-Luddite Manifesto," *NASTS News*, May 1990.

18. J. Ellul, *The Technological Society* (New York: Alfred A. Knopf, 1964).

19. L. Mumford, *The Myth and the Machine: Technics and Human Development* (New York: Harcourt Brace Jovanovich, 1967); *The Myth and the Machine: The Pentagon of Power* (New York: Harcourt Brace Jovanovich, 1970).

20. T. Roszak, *Where the Wasteland Ends: Politics and Transcendence in Post-Industrial Society* (New York: Doubleday Anchor Books, 1973), 239.

21. P. Feyerabend, *Science in a Free Society* (London: New Left Books, 1978).

22. Winner, *Autonomous Technology*.

23. J. Turner, "Democratizing Science: A Humble Proposal," *Science, Technology & Human Values* 15, no. 3 (1990): 336–359.

24. S. Woolgar, *Science: The Very Idea* (London: Tavistock Publications, 1988).

25. B. Latour, *Science in Action* (London, Milton Keynes: Open University Press, 1987).

26. Glendinning, "Notes."

27. Gerald Holton, *Science and Anti-Science* (Cambridge, Mass.: Harvard University Press, 1993), 148.

28. P. R. Gross and N. Levitt, *Higher Superstition: The Academic Left and Its Quarrels with Science* (Baltimore: The Johns Hopkins University Press, 1994).

29. M. Zimmerman, "Newspaper Editors and the Creation-Evolution Controversy," *Skeptical Inquirer* 13 (1990): 182–195.

30. *Skeptical Inquirer*, pers. comm.

▲ **CHAPTER 6.** *Recent Approaches to "Scientific Literacy"*

1. ESS: Elementary Science Study; SAPA: Science—A Process Approach; SCIS: Science Curriculum Improvement Study; COPES: Conceptually Oriented Program in Elementary Science.

2. P. De H. Hurd, "Science Literacy: Its Meaning for American Schools," *Educational Leadership* 16 (1958): 13–16; R. C. McCurdy, "Toward a Population Literate in Science," *Science Teacher* 25 (1958): 366–368.

3. D. P. Ausubel, "Some Psychological Considerations in the Objectives and Design of an Elementary School Science Program," *Science Education* 47 (1963): 278–284.

4. A. S. Fischler, "Science, Process, the Learner: A Synthesis," *Science Education* 49, no. 3 (1965): 402–409.

5. J. M. Atkin, "'Process' and 'Content' in Grade Schools," *Science* 151 (1966): 1033.

6. National Science Board Commission, *Educating Americans for the 21st Century* (CPCE-NSF-03) (Washington, D.C.: National Science Foundation).

7. M. H. Shamos, "Views of Scientific Literacy in Elementary School Science Programs: Past, Present, and Future," in *Scientific Literacy*, ed. A. Champagne et al. (Washington, D.C.: AAAS, 1989).

8. T. Bredderman, "Effects of Activity-Based Elementary Science on Student Outcomes: A Quantitative Synthesis," *Review of Educational Research* 53, no. 4 (1983): 499–518.

9. W. W. Welch, "Twenty Years of Science Curriculum Development: A Look Back," in *Review of Research in Education*, ed. D. C. Berliner (Washington, D.C.: American Educational Research Association, 1979).

10. SRI International, *Opportunities for Strategic Investment in K–12 Science Education: Options for the National Science Foundation* (Menlo Park, Calif.: SRI Project No. 1809).

11. Office of Technology Assessment, *Elementary and Secondary Education for Science and Engineering* (OTA-TM-SET-41) (Washington, D.C.: U.S. Government Printing Office, 1988).

12. M. Shamos, "STS: A Time for Caution," in *The Science, Technology, Society Movement*, ed. R. Yager (Washington, D.C.: National Science Teachers Association, 1993).

13. Shen, "Scientific Literacy and the Public Understanding of Science."

14. K. Keniston at Cornell STS Symposium, as reported by W. Lepkowski, *Chem. Eng. News*, March 20, 1989, 41.

15. D. Nelkin at Cornell STS Symposium.

16. R. Weirich, "Science, Technology and Society (STS): Its Meaning and an Assessment of Eighth and Eleventh Grade Maine Students' Knowledge of STS and Selected Factors that Affect STS Knowledge" (Ed.D. dissertation, University of Maine, Orono, Maine, 1988).

17. P. Rubba and R. Wiesenmayer, "A Goal Structure for Precollege STS Education: A Proposal Based upon Recent Literature in Environmental Education," *Bulletin of Science, Technology and Society* 5, no. 6 (1985): 573–580.

18. P. Rubba, "Recommended Competencies for STS Education in Grades 7–12," *High School Journal* 70, no. 3 (1987): 145–150.

19. S. Cutcliffe, "Science, Technology, and Society Studies as an Interdisciplinary Academic Field," *Technology in Society* 11, no. 4 (1989): 419–425; idem, "The STS Curriculum: What Have We Learned in Twenty Years?" *Technology in Society* 15, no. 3 (1990): 360–372.

20. R. E. Yager and R. Roy, "STS: Most Pervasive and Most Radical of Reform Approaches to Science Education," in *The Science, Technology, Society Movement*, ed. R. E. Yager (Washington, D.C.: National Science Teachers Association, 1993), 7–13.

21. M. H. Shamos, "STS: A Time for Caution," in *The Science, Technology, Society Movement*, ed. R. E. Yager (Washington, D.C.: National Science Teachers Association, 1993), 65–72.

22. L. R. Graham, *Science in Russia and the Soviet Union* (New York: Cambridge University Press, 1993).

23. R. Yeany, "A Unifying Theme in Science Education?" *NARST News* 33, no. 2 (1991): 1–3.

24. B. Reeves and C. Ney, "Positivist and Constructivist Understandings about Science and Their Implications for STS Teaching and Learning," *Bulletin of Science, Technology and Society* 12 (1992): 195–199.

25. *NSTA Reports!*, February/March 1993, 2.

26. American Association for the Advancement of Science, "Science for All Americans," AAAS Publications 89-01S, 89-02S, 89-03S, 89-04S, 89-05S, 89-06S (Washington, D.C., 1989). The year "2061" is the year in which Halley's comet will next return.

27. American Association for the Advancement of Science, *Benchmarks for Science Literacy* (New York: Oxford University Press, 1993).

28. National Science Teachers Association, Washington, D.C.

29. As quoted in the *Daily Telegraph*, British Association Extra, September 1989.

30. H. Spencer, *Essays on Education, Etc.* (London: J. M. Dent and Sons, Ltd., 1911), 3–44.

31. B. Chapman, "The Overselling of Science in the Eighties," *School Science Review* (Brit.) 72, no. 259 (1991): 9–35.

32. C. Ailes and F. Rushing, "Soviet Math and Science Educational Reforms During *Perestroika*," *Technology in Society* 13 (1991): 109–122.

33. Available through the Department of Technology and Society, SUNY at Stony Brook, New York, NY 11794-2250.

34. See S. Goldberg, ed., *The New Liberal Arts Program: A 1990 Report* (Stony Brook, N.Y.: NLA Center, Department of Technology and Society, State University of New York, 1990).

▲ **CHAPTER 7.** *The Future of "Scientific Literacy"*

1. R. M. Hazen and J. Trefil, *Science Matters: Achieving Scientific Literacy* (New York: Doubleday, 1991).

2. Hirsch, Kett, and Trefil, *The Dictionary of Cultural Literacy*.

3. E. L. Boyer, *High School: A Report on Secondary Education in America* (New York: Harper and Row, 1983), 132–134.

4. E.g., the National Research Council, through its Committee for Science Education Standards and Assessment.

5. E. L. Boyer, *High School Report*, 179.

6. W. Aldridge, "Why NSTA Should Certify Science Teachers," *Science Teacher*, December 1984, 20.

7. T. Bell, *The Thirteenth Man* (New York: Free Press, 1988), 180.

8. S. Tobias, *They're Not Dumb, They're Different: Stalking the Second Tier*

(Tucson: Research Corporation, 1990); "Math Anxiety and Physics: Some Thoughts on Teaching 'Difficult' Subjects," *Physics Today* 38, no. 6 (1985): 61–68; "Peer Perspectives on the Teaching of Science," *Change* 18 (March/April 1986): 36–41; "Peer Perspectives on Physics," *Physics Teacher* 26 (1988): 77–80; S. Tobias and R. R. Hake, "Professors as Physics Students: What Can They Teach Us?" *Am. J. Physics* 56 (1988): 786–794.

9. A. Emmett, "Arrogance, Poverty, and Hierarchy Are Hidden Turnoffs in Science Education," *The Scientist* 5, no. 6 (1991): 11.

10. S. Tobias, *Revitalizing Undergraduate Science: Why Some Things Work and Most Don't* (Tucson: Research Corporation, 1992).

11. G. W. Tressel, "Public Understanding and the School Curriculum," Columbia University Scientific Literacy Seminar, October 23, 1990. As reported in *Teachers Clearinghouse for Science and Society Newsletter* (New York), Winter 1991.

▲ **CHAPTER 8.** *How Much Scientific Literacy Is Really Needed?*

1. The AAAS Project 2061 seeks to address this issue through its book *Benchmarks for Scientific Literacy*, which specifies levels of understanding and abilities by grade level that students are expected to reach as they move on to becoming "literate "in science (New York: Oxford University Press, 1993).

2. M. H. Shamos, "A False Alarm in Science Education," *Issues in Science and Technology* 4, no. 3 (1988): 65–69.

3. M. H. Shamos, "The Lesson Every Child Need Not Learn," *The Sciences*, July/August 1988, 14–20.

4. See, e.g., *Science* 256 (1992): 172; *Chemical & Engineering News*, April 20, 1992, 14.

5. J. D. Miller, "Scientifically Illiterate (Study of Adult Understanding of Terms and Concepts)," *American Demographics* 9 (1987): 26; idem, "Scientific Literacy," speech delivered at annual meeting of the AAAS, San Francisco, January 1989.

6. Mark Twain, *Roughing It*, chapter 48 (1872).

7. E. Teller, at Hearings, Senate Armed Services Committee, November 25, 1957.

8. J. B. Conant, ed., *Harvard Case Histories in Experimental Science* (Cambridge, Mass.: Harvard University Press, 1948).

9. H. Poincaré, *The Value of Science*, trans. by George B. Halstead (New York: Science Press, 1907), 8.

10. I. I. Rabi, in a lecture presented to a meeting of the Joseph Priestley Association, New York, November 1, 1986.

11. J. B. Conant, *Foreword*, in I. B. Cohen and F. G. Watson, eds., *General Education in Science* (Cambridge, Mass.: Harvard University Press, 1952).

12. See, e.g., W. D. Rifkin, "Whom to Heed in the Expert Society," *Bulletin of Science, Technology and Society* 10 (1990): 156–160; see also S. Jasanoff, *The Fifth Branch* (Cambridge, Mass.: Harvard University Press, 1990).

13. J. Ellul, *The Technological Society* (New York: Alfred A. Knopf, 1964).

14. L. Mumford, *The Myth and the Machine: Technics and Human Development* (New York: Harcourt Brace Jovanovich, 1967); idem, *The Myth and the Machine: The Pentagon of Power* (New York: Harcourt Brace Jovanovich, 1970).

15. A. Toffler, *Future Shock* (New York: Random House, 1970).

16. T. Roszak, *Where the Wasteland Ends: Politics and Transcendence in Post-Industrial Society* (New York: Doubleday Anchor Books, 1973), 239.

17. P. Feyerabend, *Science in a Free Society* (London: New Left Books, 1978), 88–89.

18. Learned Hand, speech delivered at the Convocation of the Board of Regents, University of the State of New York, October 24, 1952.

19. R. Hoffman, "Why Scientists Shouldn't Run the World," *Issues in Science and Technology* 7, no. 2 (Winter 1990–91): 38–39.

20. C. Frankel, "The Nature and Sources of Irrationalism," *Science* 180 (1973): 927–931.

21. A. Kantrowitz, "Proposal for an Institution for Scientific Judgment," *Science* 156 (1967): 763–764; idem, "Controlling Technology Democratically," *American Scientist* 63 (1975): 505–509.

22. See, e.g., P. W. Huber, *Galileo's Revenge: Junk Science in the Courtroom* (New York: Basic Books, 1991).

23. D. Nelkin, "Thoughts on the Proposed Science Court," *Newsletter on Science, Technology and Human Values* 18 (1977): 20–31.

24. S. Jasanoff and D. Nelkin, "Science, Technology and the Limits of Judicial Competence," *Science*, December 11, 1981, 196–200.

25. Jasanoff and Nelkin, "Limits of Judicial Competence."

26. A. Kantrowitz, pers. comm.

27. "Science and Technology in Judicial Decision Making: Creating Opportunities and Meeting Challenges" (New York: Carnegie Corporation, 1993).

28. T. Roszak, *The Making of a Counter-Culture* (Garden City, N.Y.: Doubleday, 1969).

29. R. A. Schibeci, "Public Knowledge and Perceptions of Science and Technology," *Bulletin of Science, Technology and Society* 10 (1990): 86–92.

▲ CHAPTER 9. *Conclusions*

1. W. D. Rifkin, "Whom to Heed in the Expert Society," *Bulletin of Science, Technology and Society* 10 (1990): 156–160.

2. Center for Unified Science, Victor Showalter, Director, Headquartered at Capital University, Columbus, Ohio.
3. M. Perutz, *Is Science Necessary?* (New York: E. P. Dutton, 1989).
4. J. B. Conant, ed., *Harvard Case Histories in Experimental Science* (Cambridge, Mass.: Harvard University Press, 1948).

▲ **EPILOGUE**

1. See preface and chapter 7.
2. Source: U.S. National Center for Education Statistics, *Digest of Education Statistics*, Annual.
3. Alice Rossi, "Women in Science: Why so Few?" *Science* 148 (1965): 1196–1202.
4. J. D. Stern, "The Condition of Education: A Statistical Report" (Washington, D.C.: U.S. Department of Education, Office of Educational Research and Improvement, 1986).

AAAS. *See* American Association for the Advancement of Science
adult education courses, 186–188
advanced placement courses, 10
advisory bodies, 209
Age Discrimination in Employment Act, 240n15
Ailes, C., 154
Aldridge, W., 248n6
ALF. *See* Animal Liberation Front
Alfred P. Sloan Foundation, 155
"alphabet soup" programs, 130, 218, 246n1
alternative medicine, 112
American Association for the Advancement of Science (AAAS), 87; Project 2061, 150–152, 237, 248n26, 249n1
Animal Enterprise Protection Act (1992), 122
Animal Liberation Front (ALF), 120, 122
animal research, 121
animal rights movement, 120–122
animal welfare movement, 120
anti-intellectualism, 116
anti-science movement, 114–120
anti-technologists, 114–120
applied science, 39. *See also* technology

"Aquarian conspiracy," 113
Archimedes, 26
Aristotle, 25–28
Arnold, Matthew, 34
Astin, A., 245n30
Astin, H., 245n30
astrology, 55, 116
astronomy, 62
Atkin, J. M., 246n5
Atkinson, Richard, 12
atom bomb, 76, 142
Australia, education in, 211–212
Ausubel, D. P., 246n3
"autonomous technology" (Winner), 116

Bacon, Sir Francis, 29, 56
basic science, vs. technology, 109–110. *See also* pure science
Bell, Terrel, 176
Berger, J. O., 243n28
bias, 211, 213–214
"big science," support for, 109
biology, 95–96, 105–106
biomedical sciences, 68, 112, 121
biotechnology industry, 123
Bloom, Allan, 101–102
Boorstin, Daniel, 60, 76
Born, Max, 57
Boyer, Ernest, 175

Boyle, Robert, 62–63
Bredderman, T., 133, 134
Bridgman, Percy, 64
Britain. *See* United Kingdom
Bronowski, Jacob, 27, 50, 103–104
budgets. *See* funding
Bulletin of Atomic Scientists [journal], 76
Burnham, John, 36, 37, 84–85
Bush, George, 164, 232
business/industry community, 15, 162, 163, 187. *See also* industry

Campbell, Norman, 47
Capra, Fritjof, 112–113
career opportunities, 10–21, 183, 235–236
Carnegie Commission, 210
Cavendish, Henry, 62
certification programs, for teachers, 176
Champagne, Audrey, 5, 79
Chapman, Bryan, 154
Cicero, 28
citizens' groups, 208
"civic scientific literacy" (Shen), 140, 216–217
Civil War, 38
Clarke, F. W., 241n28
colleges and universities, 17, 38, 106, 155, 213; admission requirements, 40, 162; curricula, 146, 155–156; distribution requirements, 109, 184; education departments, 176; graduation requirements, 162–163, 164; public and private, 234; technical, 37
college students, 124, 196
Committee for the Scientific Investigation of the Paranormal, 246n16
Committee of Ten. *See* National Education Association
common sense, 57–60, 168, 209
Conant, James B., 52, 86, 199, 201, 225
conceptual schemes, 46, 67, 89, 95. *See also* theories

Conceptually Oriented Program in Elementary Science (COPES), 130, 173, 246n1
conflict resolution, 209
constructivism, 117, 148–149
COPES. *See* Conceptually Oriented Program in Elementary Science
"corrigible fallibility" (Nash), 52
cost-benefit analysis, 172
counterculturists, 203–205
courts.*See* litigation
Cranberg, Lawrence, 54
creation-evolution controversy, 125–126
critical thinking, 81, 90, 96, 122, 226
"cultural literacy" (Hirsch), 87
cultures. *See* two-cultures controversy
curriculum development projects, 83, 129, 136–139
curriculum reform, 1, 83–84, 129–161, 218–219, 230, 248n26; elementary, 129–132; evaluation of, 169
Cutcliffe, Stephen, 145

Dampier, W., 240n3
Daniels, George, 38
Davis, I. C., 81
deduction, 25
demarcation problem, 53
democracy, 103–104, 197; and technology, 203. *See also* science, and democracy; science, and society; technology, and society
Democritus, 242n5
Dewey, John, xi, 1, 77–80
Draize eye irritancy test, 121

Earth Observing System, 93–94
Eastern philosophy, 113
Eaton, Amos, 37–38
Eddington, A. S., 242n7
education: future of, 232; support for, 174. *See also* science education
educational reform, 232

educators, 160; as elitists, 180; scientists as, 130, 137, 187. *See also* mentors; teachers
Einstein, Albert, 110, 168
Eisenhower, Dwight D., 38
elementary schools, 129, 177–179. *See also* science education, in elementary school
Elementary Science Study (ESS), 130, 133–134, 173, 246n1
Ellul, Jacques, 115, 203
employment in science fields. *See* career opportunities; job opportunities; work force, scientific
engineering, 71, 236
environmental issues, 167, 226
ESS. *See* Elementary Science Study
ethics, 187
Euclid, 26
evaluations: of educational programs, 169–171; of science curricula, 171–172; of testing, 233
evolution-creation controversy, 125–126
examinations, 161–164, 170–171, 233
expert opinion, 201–204, 211, 217
explanation, in science, 26–28, 46

facts and values, 208
falsifiability, 53–54
Faraday, Michael, 30
fear of science, 118–119
Fechter, Alan, 11
Federation of American Scientists, 76
Federation of Atomic Scientists, 76
Ferguson, Marilyn, 112–113
Feyerabend, Paul, 116, 204
Fischler, A. S., 246n4
foreign students, 18
Foundation on Economic Trends, 122–125
Frankel, Charles, 205
Franklin, Benjamin, 35–36
Freedom space station, 93–94
funding: agencies, 173; for "big science" projects, 93; for education, 178, 233

Galileo Galilei, 29, 64
Galileo spacecraft, 123
Galvani, Luigi, 30
Geller, Uri, 112
gender imbalance, in science professions, 235–236
Gingerich, O., 242n11
Glendinning, C., 119, 246n17
"Goals 2000" program, 233
government: certification programs, 176; influence, 163–164; support, of science and technology, 38–39, 83, 92–93, 163–164, 232–234
government agencies, 83, 211. *See also individual names*
Graham, L. R., 248n22
Great Britain. *See* United Kingdom
"great ideas" in science, 159
Greece, Golden Age of, 24, 66
Greenberg, Daniel, 76
Greene, J. C., 36
Gross, P. R., 119

Haldane, J.B.S., 35
hands-on activities, in science education, 130–133, 137, 169
Harman, David, 97
Harvard Case Histories in Experimental Science, 199
"Harvard Project Physics," 134
Hazen, R. M., 159
health care, 112
Hechinger, Fred, 3
Heininger, S. Allen, 11
Higher Education Research Institute (HERI), 96
high schools, 40–41; curricula, 40, 139, 150–155, 181; graduation requirements, 162–163, 164, 181
high school students, 139; as mentors, 179
Hirsch, E. D., 87, 88, 101–102
Hoff, A.S.G., 244n17
Hoffman, Roald, 205

Hofstadter, Douglas, 58–59
Holton, Gerald, 119
Huber, P. W., 250n22
Human Genome Project, 92–93
Hume, David, 57, 243n32
Hurd, Paul, 131
Huxley, Aldous, 103
Huxley, Thomas, 30, 33–34, 57, 59

illiteracy, 97
imbalance: of gender, 235–236; of
    two cultures, 107
incentives to learn science, 162, 183,
    196. *See also* motivation
induction, 27
"industrial arts," 68
Industrial Revolution, 67
industry: influence on education of,
    121, 163; job market in, 13–17,
    131, 162; support from, 15, 187; in
    United Kingdom, 163
informal education, 184–186
innumeracy, 95, 96
inquiry-based science programs, 137
in-service teacher training, 135, 137,
    177
Institution of Civil Engineers, London,
    71
"integrated science" movement, 219
irrationalists vs. technology, 206
irrationality, 203–205

Jasanoff, Sheila, 208
Jefferson, Thomas, 35
Jefferys, W. H., 243n28
job market: for nonprofessionals, 98–
    99; for science professionals, 13–
    17, 131, 162
job opportunities, 11, 162
"junk science," 208, 210, 250n22
juries, 195, 204. *See also* litigation

Kant, Immanuel, 244n12
Kantrowitz, Arthur, 207, 209, 213
Keniston, Kenneth, 143
Kett, J. F., 88

Klopfer, L. E., 79
knowledge revolution, 113

laboratory animals, 120–121
labor force, general, 98. *See also* job
    market
Latour, Bruno, 117
laws and theories, 62, 242n10
lawsuits. *See* litigation
Leavis, F. R., 107
lectures, public, 37
Leonardo da Vinci, 28
Levitt, N., 119
libraries, public, 186–187
Lightman, A., 242n11
litigation, on scientific issues, 124,
    207–210
Locke, John, 243n32
logic, 24–25
Luddism, 114

McCurdy, Richard, 131
Machiavelli, Niccolò, 69
MacLaine, Shirley, 113
MacVicar, M.L.A., 240n16
mass media, 31, 89, 125–127, 185–
    186, 202, 213
mathematics, 64–65, 95–96, 226,
    236
medicine, 112
mentors, in elementary science pro-
    grams, 179
methodology, 56
military use of science, 77
Miller, Jon, 81, 87, 88, 90
motivation, of students and teachers,
    151, 230
Mumford, Lewis, 115, 203

Nagel, Ernest, 47, 58
Napoleon I, 30
NASA. *See* National Aeronautics and
    Space Administration
Nash, L. K., 52
national certification program, for
    science teachers, 176

national curriculum, 163; standards for, 149, 164, 176, 233; and testing, 164

National Academy of Sciences, 209, 233

National Aeronautics and Space Administration (NASA), 93–94, 123

National Defense Education Act (NDEA) (1958), 83

National Education Association (NEA), 40

National Research Council (NRC), 11, 149, 248n4

National Science Board, 2, 133

National Science Foundation (NSF), 4, 10–11, 39, 83, 84, 135, 136, 153, 173, 185, 232

National Science Teachers Association (NSTA): certification program, 176; Scope, Sequence and Coordination (SS&C) Project, 152–153, 237

natural history, 26, 56, 66, 96, 219–220

nature, theories of, 24, 46, 62, 112

NDEA. *See* National Defense Education Act

NEA. *See* National Education Association

Nelkin, Dorothy, 143, 208

neo-Luddites, 106, 114, 118, 124

"New Age" movement, 112

"New Liberal Arts," 155–156

newspapers, 3, 125–127, 185–186

new technologies, 114, 123, 124

Newton, Sir Isaac, 30, 61, 243n29

Ney, C., 149

NIMBY (Not In My Backyard) factor, 140–141

Noll, V. H., 244n18

NRC. *See* National Research Council

NSF. *See* National Science Foundation

NSTA. *See* National Science Teachers Association

null experiments, 53

numeracy, 156. *See also* innumeracy; quantitative reasoning

Occam, William of, 243n27

Occam's Razor, 61

Office of Science and Technology Policy, U.S., 38–39

Oppenheimer, Robert, 76

Paulos, J. A., 245n29

Pearson, Karl, 56, 79, 243n32

People for the Ethical Treatment of Animals (PETA), 120–121

Perutz, Max, 223

PETA. *See* People for the Ethical Treatment of Animals

Ph.D. supply, 19–20

Philadelphia Academy of Science, 35

physics, 113; careers in, 236

pilot studies, of educational reforms, 172

pipeline, 182–183. *See also* career opportunities; job market; science education; work force, scientific

Plato, 23

Poincaré, Henri, 35, 60, 62

popular press, 3, 89, 125–127, 186, 202, 213. *See also* mass media; newspapers

Porter, Sir George, 153

positivism, rejection of, 149

postmodernism, 117, 148–149

President's Science Advisory Committee (PSAC), 38

Prewitt, Kenneth, 85

Priestley, Joseph, 55

primary grades. *See* elementary schools

private schools, 234–235

privatization of public schools, 234–235

professionalism, in education, 174–176

professional societies, scientific, 37

proficiency, in science, 163

progress, 52

Project 2061 (AAAS), 150–152, 237, 248n26, 249n1

PSAC. *See* President's Science Advisory Committee

pseudoscience, 54–55, 113, 127, 202, 203, 246n16
PSSC physics curriculum, 83
public schools, privatization of, 234–235
"public good," 123, 127
public interest groups, 76, 211
public relations, 211
"publics," defined, xii
publishers, and curriculum reform, 136, 158
Publisher's Initiatives Program, 136
pure science, 39, 226

quantitative reasoning (numeracy), 156, 165

Rabi, I. I., 200
rationalism, rejection of, 115–116, 148–149
rationality, 205
reason: appeals to, 205–206; vs. emotion, 122; rejection of, 112–113, 115–116
reasoning skill, 166
Reeves, B., 149
Reich, Robert, 15
Reid, Thomas, 58
relativity theory, 168
relevance of science knowledge, 226
research: with animals, 121; support for, 34–35, 110, 147
Rifkin, Jeremy, 122–125
Rifkin, William D., 217, 250n12
risk-benefit analysis, 69–70, 122–123
Roentgen, Wilhelm Conrad, 110
role models, in education, 108
Rossi, Alice, 236
Roszak, Theodore, 115, 204
Roy, R., 247n20
Royal Institution, London, 30
Royal Society of London, 71
Rubba, P., 144–145
Rushing, F., 154
Russell, Bertrand, 242n4
Russia, 152
Rutherford, Ernest, 110

SAPA. *See* Science—A Process Approach
Schibeci, R. A., 211–212
schools, U.S., 174. *See also* colleges and universities; elementary schools; high schools
schools of education, 175–176
school systems, local and state, 232
science: advisory bodies, 209; appreciation, as educational goal, 159, 197–199; contemporary relevance of, 226; and controversy, 207; and counterculture movement, 111–127, 148–149; criticism of, 111, 112–113, 118–122, 147–148; and culture, 102–127; and democracy, 103–104, 116–117, 197; descriptive, 95–96; disciplines, 218; educators, 137, 160; and the environment, 110; essentials, 151; fear of, 118–122; history of, 25–31, 61–63, 66, 199, 225; and humanities, 101–127; ignorance of, 106; image of, 91, 125, 180, 211–212, 219, 221; interest in, 187, 197, 224; and mathematics, 65, 95–96; medical, 68, 112, 121; military use of, 77; and moral/ethical problems, 123; museums, 187; and nature, 46; nature of, 46–48, 94, 141–142, 180, 218–219, 221, 223–226; news, 125; and opinion, 168; partitioning of, 218; process of, 132, 134, 137, 218; and the "public good," 123; public support for, 197, 200, 216; public understanding of, 86, 141, 159–161, 171, 185, 190, 200, 211, 216, 224; relevance of 165–169; rigidity of, 168; "shows," 30, 37–38; and society, 82, 85, 91–94, 117, 123, 139–150, 166–168, 189–197, 200–201, 206–214; specialists, in public schools, 178; student involvement in, 168–169; and technology, 24, 66–69, 71, 91, 109–110, 116, 224, 230; value-free, 213–214; watch

committees, 212–214. *See also* social issues, science/technology-based; *particular disciplines (e.g.,* astronomy; biology)

Science—A Process Approach (SAPA), 130, 133–134, 173, 246n1

science-based vocabulary, 87, 88

science-bound students, 12, 181–183

"Science Court" (Kantrowitz), 207–210

science curricula, 219, 223–228; college level, 155–156; elementary, 173; funding for, 173; high school, 40, 139, 150–155, 181; nature of, 146; revision of, 173; role of, 230–232

Science Curriculum Improvement Study (SCIS), 130, 133–134, 173, 246n1

science education, xi, xiii–xiv, 2, 129–156, 160, 218–219; assessment of, 172; college level, 155–156, 184; compulsory, 191; in elementary school, 129–139, 177–179; evaluation of, 171–172; future of, 157–159, 165; goals of, 35–38, 216–217, 231; history of, 30–43, 82–84, 102–106, 139–150, 197, 237–238; informal, 126–127, 184–188; public funding of, 178; public support for, 216; purpose of, 35–38, 73–82, 161, 183, 189, 190, 198–201, 206; reform of, 190, 196; in secondary school, 139–155, 183; in Soviet Union, 154; standards for, 233; in United Kingdom, 153–154

science/engineering pipeline, 182–183

science/technology-based social issues, 76–78, 91–94, 140–141, 146–147, 166–168, 190, 196, 202–204, 207–214, 226–227

Science/Technology/Society. *See* STS movement

scientific: awareness, as educational goal, 223; experts, 200, 206–210,

217; ideas vs. common sense, 168; illiteracy, 96, 111, 126–127; inquiry, 221; journals, 37; laws, 62; opinion, 208; societies, 71; truth, 47–48, 50–51; work force, 10–21, 190

"scientific attitude" (Dewey), 80–81

"scientific habits of the mind" (Dewey), 89, 96, 122, 200

scientific literacy, 81, 86–90, 158–163; adult, xvi, 132–133, 170, 184; cultural, 87; definition of, 84–87, 144, 150, 169, 229; in democracy, 191; as educational goal, xii, 131, 158, 177–178, 198, 216, 229; effects of, 188; extent of, 94, 191–195, 215–217; failure of efforts toward, 160; functional, 88; levels of, 87–90; meaning of, 216; measurement of, 170–171; movement, 80–82, 115, 216; need for, 98–99; purpose of, 45, 85; and STS literacy, 144–145; social, 189–190, 249n1; standards for, 189, 249n1; of students, 138–139; "true," 89

"scientific method," 56

"scientific revolution," 29, 66

"scientific savvy" (Prewitt), 85

scientists and engineers: education of, 182–183; as educators, 130, 137, 187; image of, 182, 211, 235–236; and nonscientists, 106–107; opinions of, 202–204; supply of, 11, 36, 41

SCIS. *See* Science Curriculum Improvement Study

SDI. *See* Strategic Defense Initiative

secondary schools, 40–41. *See also* high schools

self-correction, in science, 51–52, 142, 221

self-expression, by students, 168

separation of cultures, 104

sex discrimination, 236

Shamos, Morris H., 243n31, 244n24, 245n28, 247nn7, 12, 21, 249nn2, 3

Shen, Benjamin, 85, 140

simplicity, as scientific goal, 61
Simpson, George Gaylord, 47
Skeptical Inquirer, The [journal],
    246n16
Snow, C. P., 101–106
social issues, science/technology-
    based, 76–78, 91–94, 140–141,
    146–147, 166–168, 190, 196,
    202–204, 207–214, 226–227
social literacy, 77, 216–217
social scientists, 117, 146, 214, 216–
    217
social studies, 142, 146, 227
Society for Literature and Science,
    107
sociologists, 117, 205, 208
Sociolgy of Scientific Knowledge
    (SSK), 117
Socratic method, 25
Soviet Union, science in, 148, 154
Spencer, Herbert, 31–33, 106
Sputnik, 83
SSC. *See* Superconducting Super-
    collider
SSK. *See* Sociology of Scientific
    Knowledge
standards for scientific literacy, 189,
    249n1; national, 149, 164, 176,
    233
Stanford University, 217
"Star Wars," 91–92, 245n27
state education departments, 176,
    232
Stephan, Paula, 19
Stern, J. D., 251n4
Strategic Defense Initiative (SDI), 91–
    92, 245n27
"STS literacy," 144–145
STS (Science/Technology/Society)
    movement, 108, 115, 139–150,
    168, 205, 227
Superconducting Supercollider (SSC),
    92–93
surveys, 126, 211

teachers, 174–179; certification of,
    176; education of, 176–178; ele-
mentary school, 177; evaluation of,
    175; high school, 177; image of,
    180; qualifications of, 145, 176–
    177, 228; science understanding
    of, 228, 234; shortages of, 6–7; sta-
    tus of, 174–176; training of, 135–
    138, 175–178
teacher training institutions, 83, 175–
    177
technological literacy, 68, 99, 147,
    155–156
technology, 66–71, 224–227; control
    of, 123–124; criticism of, 147–
    148, 208; in daily life, 198; and de-
    mocracy, 203–204; education, 68,
    141, 156, 229, 237; effects of, 167;
    future of, 165; history of, 66–67,
    225; image of, 114–117; and
    moral/ethical problems, 123; rejec-
    tion of, 109, 118–119, 122, 206;
    research in, 34; and risk, 122; and
    science, 109, 230; and society, 123,
    142, 147, 201, 220–221; under-
    standing of, 156. *See also* science,
    and society
teleology, 27
Teller, Edward, 197
tests. *See* examinations
test scores, 8–9
theories 89–90, 95–96, 138, 141,
    222, 242n10. *See also* conceptual
    schemes
Thorburn, W. M., 243n28
Tobias, Sheila, 180–181
Toffler, Alvin, 203
Tocqueville, Alexis de, 36
Trefil, J., 88, 159
Tressel, George, 185
Triad projects, 136–139
Truman, Harry, 38
truth, in scientific endeavor, 47–48,
    50–55, 222
Turner, Joseph, 116
two-cultures controversy, 101–127

uncertainty, in science, 123, 208
Uncertainty Principle, 63n33

"understanding" of science, 99–100
unified science movement, 219
United Kingdom, 211; National Curriculum, 153, 163; science education in, 32–33
United States: Department of Education, 39, 176, 232; education in, 232–234; Office of Education, 39, 83; Office of Science and Technology Policy, 38–39; Supreme Court, 210
universal laws, 62
universities. *See* colleges and universities
U.S.S.R. *See* Soviet Union

value-free knowledge, 208–209
value judgments, 214
Volta, Alessandro, 30
Voltaire, 30

watch committees, scientific, 213
watchdogs, 124
Weirich, Richard, 144
Welch, W. W., 134
whistle-blowers, 124
White, Robert, 11
Wiesenmayer, R., 144
William of Ockham, 243n27
Winner, Langdon, 111, 116, 146
women in science and engineering professions, 236
Woolgar, Steve, 117
work force, scientific, 10–21, 190
World War II, xii, 142, 163

Yager, R. E., 247n20
Yeany, R., 248n23

Zimmerman, Michael, 125
Zuckerman, Mortimer, 42

▲ ABOUT THE AUTHOR . . .

Morris H. Shamos views the problem of scientific literacy from three different perspectives. First, he was a university-based scientist, having served for many years as professor and chair of the Physics Department in Washington Square College of New York University. Second, he has been a science educator and teacher, spanning all the grade levels from elementary school through graduate school. He served as president of the National Science Teachers Association in 1967–68, when he was concerned mainly with high school science, and in the early 1970s he directed one of the major national science curriculum development programs for elementary schools (COPES). Third, Dr. Shamos understands the literacy dilemma from the point of view of industry, having spent more than two decades as a senior executive officer and chief scientist of one of the world's leading corporations in the medical instrument field.

Morris Shamos was born in Cleveland, Ohio. He was educated in the public schools there, and completed his university and graduate studies at New York University, receiving his Ph.D. in physics in 1948. Among other organizations, Dr. Shamos is a member of the American Physical Society, American Chemical Society, American Association for Clinical Chemistry, Institute of Electronic and Electrical Engineers, American Association of Physics Teachers, National Science Teachers Association, and a fellow of the AAAS and the New York Academy of Sciences, of which he was president in 1982. He has served as a board member of several corporations and in an advisory capacity to a number of educational organizations. He is currently Professor Emeritus at New York University and directs a private consulting practice in New York City.

Dr. Shamos has published extensively in the scientific and educational literature, including authorship or editorship of three books, and continues to be active in the ongoing debate on scientific literacy for the general public.